# FUTURE EVOLUTION

# FUTURE EVOLUTION

PETER WARD

Images by
ALEXIS ROCKMAN

Foreword by
NILES ELDREDGE

*A W. H. Freeman Book*
TIMES BOOKS
Henry Holt and Company
*New York*

To H. G. Wells and his descendants

TIMES BOOKS
Henry Holt and Company, LLC
*Publishers since 1866*
115 West 18th Street
New York, New York 10011

*I would like to thank the following people for various reasons:*

| | | |
|---|---|---|
| *Sam Fleischman* | *Diana Blume* | *Rob DeSalle* |
| *Jill Rowe* | *Kirk Johnson* | *Daniel Heiminder* |
| *John Michel* | *Kurt Keifer* | *Carl Zimmer* |
| *Niles Eldredge* | *Andrew Vallely* | *Alisa Tager* |
| *Jean-Jacques Annaud* | *Tom Sanford* | *Anne Pasternak* |

*And I would like to thank my gallerists Jay Gorney, Karin Bravin, and John Lee, and especially Rodney Hill.*—A.R.



Library of Congress Cataloging-in-Publication Data

Ward, Peter
Future evolution / Peter Ward; images by Alexis Rockman; foreword by Niles Eldredge.
p. cm.
Includes bibliographical references (p. ).
ISBN 0-7167-3496-6 (cloth)
1. Evolution (Biology) I. Title.
QH366.2 .W37 2001
576.8—dc21 2001003607

Henry Holt books are available for special promotions and premiums.
For details contact: Director, Special Markets

First Edition 2001

Designed by Diana Blume

Printed in Hong Kong

10 9 8 7 6 5 4 3 2 1

# CONTENTS

# IMAGES

# IMAGES



*These drawings were created with pigments derived from the medium in which the actual fossils were discovered, provided to the artist by the author.

All images courtesy of Gorney Bravin + Lee, New York. All works were photographed by Oren Slor, New York.

FOREWORD

# BIOLOGICAL FUTURES


*Niles Eldredge*

Committee on Evolutionary Processes, and Division of Paleontology,
The American Museum of Natural History,
Central Park West at 79th Street, New York, New York 10024


Predict the future? The future of evolution of life on earth–human life, bird life, fungal life? Most of us New Yorkers would say, "Fageddabout it!" We can't even predict weather with any accuracy more than three days in advance, even with all our global monitoring stations, constant satellite imagery and computer modeling. We go to bed at night, sure that the sun will "rise," but with no way of knowing if the Dow Jones Industrial Average will rise, stay flat, or plummet. Sure, we can explain what happens ex post facto ("the snowstorm didn't materialize because high pressure from Canada deflected the low off the Jersey shore," or "the market fell on disappointing earnings reports in the tech sector"). But predicting such complex systems with any consistency remains an elusive goal, maybe even a fantasy.

So how on earth can we expect to do any better with the future of life—with its millions of species, its myriad ecosystems—projecting what's going to happen, not this week or next month, but hundreds, thousands, *millions* of years down the road? Fageddabout it!

But hold on. One of the key ingredients of the scientific endeavor, after all, is predictivity: if an idea is true, we reason, there must be certain observable consequences—and if we consistently fail to observe the predicted results, there must be something wrong with the idea itself: we must "reject the hypothesis." Laboratory experiments—though sometimes done blindly ("let's see what happens if we mix these two chemicals!")—are nonetheless usually performed with some expectation in mind of what the results will be: future outcomes, in other words, are predicted.

But even here we encounter difficulties: creationists are fond of pointing out that evolutionary biologists have usually been reluctant to predict what will happen in the evolutionary future, and claim that this failure to render testable predictions of life's future means that evolutionary biology is therefore not true science. In any event, none of us will live long enough to see if our predictions turn out to be correct.

Not so, say the philosophers: the future state of a system is not what is necessarily meant by predictivity in science; rather, for an idea to be scientific, we must simply make predictions about what we would expect to observe in the natural world if that idea is true. Thus the grand prediction of evolution would be that, if all life has descended ("with modification," as Darwin himself put it) from a single ancestor, diversifying into separate lineages as the process went along, there should be a single pattern of resemblance linking up all life on earth. More closely related species should look more like each other than more remote kin—but there should be some vestige of common inheritance of features that are found in absolutely all forms of life. That's in fact what we do see: RNA is present in all life forms; all vertebrate animals have backbones (the very meaning of the group's name), all mammals have hair.

Then, too, we would predict, were evolution "true," that in the history of life the simplest forms would have appeared first, the more complex later. That, too, we see, haunting our diagrams of the structure of relationships in the living world, but especially in the sequence of life preserved in the fossil record. Life was nothing but bacteria for its first billion or so years, and nothing but single-celled organisms for its first 2 billion years. Simpler forms of animal life preceded the more complex—and reptiles preceded their famous derivatives, birds and mammals.

So, not only is evolution a legitimately scientific concept, it is also almost certainly true—having had its two grand predictions about what life should look like (both now and in the fossil record) "corroborated" so many times over that there is no residual rational doubt that life as we know it is the product of evolution.

But if we need not predict the future for us to see the scientific nature of the very idea of evolution, is that all we can do? What about those of us who do not want to say "fageddabout it!" when we wonder what the future holds—especially for ourselves, the human beings who have so recently, so thoroughly changed the face of our globe? Is there nothing rational we can say, nothing about what we've learned from life's history that can serve as a basis for reading the future—or at least narrowing down the possibilities?

The exciting answer is "Yes!"—there's a whole lot we can say, and with confidence. Life's evolution, though usually portrayed as a string of events that saw the eventual emergence of leopards and hippos through a long lineage stretching back to primordial bacteria, can also be read as a series of patterns repeated so often that we can be sure they will happen again. Specifically, patterns of extinction followed by the appearance of new species have demonstrated without any lingering doubt that nothing much happens in the way of evolutionary change unless environmental disruption shakes life up.

Life can absorb small local disturbances (like fires or tidal waves), and replace its local dead with recruits of the same species brought in from undisturbed, neighboring ecosystems. But larger-scale disturbances—like global climate change, or objects from outer space colliding with the earth—are another matter entirely. Such larger-scale perturbations are often sufficiently devastating that they drive entire species extinct. Sometimes, as in the great mass extinctions portrayed so vividly in the pages of this book, they drive entire families, orders or even classes of animals, plants and microbes extinct. And that's when evolution kicks in.

Peter Ward, experienced paleontologist that he is, knows all this. He also sees that we are in the midst of another major surge of extinction that is bound to trigger an evolutionary rebound (indeed, he thinks it already has!). I totally agree with him that humans are the root cause of this current spasm of extinction—and that, critically, it is the fate of humanity that will determine to a very large degree the future complexion and composition of life on earth.

So the question becomes: What is our fate? Others who have taken a shot at what life's future evolution will look like have assumed that the current vector of extinction—meaning ourselves—will disappear completely, leaving all other forms of surviving life to reclaim the planet. The paleontologist and artist Dougal Dixon, in his *After Man*, made this assumption—and produced a whimsical work of great charm with his imaginary "sand sharks" and other beasts of the future.

Alexis Rockman, whose often brightly colored canvasses just as often project dark themes, has in the course of his career created a vision of life on earth totally informed by the presence of humanity—tin cans and discarded tires forming the substrate of an exuberant, ongoing Life. Alexis's vision dovetails perfectly with Peter's notion that humanity is going to survive—and that all life's future will revolve around our presence. The future is already here, with domestication of barnyard animals and the heady days of genetic engineering beginning to unfold before our eyes. A rational supposition—and one very different from Dixon's bucolic outlook.

I wonder if past cultural extinctions, where technologically advanced and complexly organized societies have disappeared even while their descendants have persisted, living simpler lives, might not also be a source of predicting the future. The current wave of human planetary disruption might cause, not our *physical* extinction so much as a loss of the "high culture,"—our knowledge—if we do overrun our Malthusian limits. Loss of topsoil, lack of access to fresh water, loss of fisheries, spread of famine, warfare and disease—all the usual apocalyptic visions, all duly acknowledged in these pages—may not drive our *bodies* extinct, but could very well play hob with our minds, our cultural memories, our *knowledge*.

It is the exercise itself that is important: the careful development of a view of the future based on what we've seen happen in the past, and what is going on right now with the human joker-in-the deck, the wild card that is mimicking the effects of the asteroid that wiped out the dinosaurs 65 million years ago. It is up to all of us to contemplate the effects of our collective lives on the future of life on our planet. Peter and Alexis have combined to develop a stimulating vade mecum, an invitation to each of us to journey along with them as we wrestle with the problem ourselves. And that, of course, is the real point about reading anything.

# PREFACE

The ringing phone was one more interruption in a day filled with them, the day just another in the blur of time we call life. I anticipated the mundane. But the telephone's slight electronic hesitation indicated *International,* and the caller's precise Oxbridge accent confirmed England. In this day of E-mail, no one pays for a phone call unless there is a pitch or a catch, so I listened with alacrity.

The smooth-spoken man asked for *Professor* Ward, showing that European formality so charming and now so extinct in America. After I confirmed that I was I, he launched into the pitch (the catch was yet to come). He explained that he was a producer for the BBC, in charge of a thirteen-hour series about evolution, then early in the planning stages. He wondered if he could ask me a few questions. Sure, I answered, using a word I am sure has never been uttered by a Brit. He explained that the series was about the future—the future of evolution, in fact. I found myself listening much more carefully, having nearly finished the manuscript for this book at the time. He reiterated that he was funded to produce thirteen *hours* of programming dealing with the animals and plants of the near to far future, with each hour profiling a future time slice, starting in the next few millennia and ending in the far-off future, a billion years from now, when the sun would be brightening to the point of threatening the existence of all life on Earth. We talked a bit further, while I kept exclaiming to myself, "Thirteen hours! What can they possibly put on the screen for thirteen hours?"

And then it was my turn to talk. I explained the thoughts that make up the subject of this book, starting out with the basic assumption that colors all that follows: for the biological life span of the planet, humanity is essentially extinction-proof, and, if we manage to develop effective interstellar travel, *completely* extinction-proof as long as the galaxy survives. Therefore, any scenario envisioning the future of biotic evolution must do so in a world dominated by humanity—just as our world is today. In such a world the range of possibilities—in particular, the probability of exotic new body plans and life forms that would make good television—is severely limited. Those that do arise will probably be small in size, for humanity has carved Planet Earth into a great number of tiny biological islands with our cities, farms, roads, and clear-cuts. Species that arise or evolve on islands usually tend to be small. In other words, there will be no new and exotic large mammals, birds, or reptiles.

From the silence that followed I gathered that I had not delivered the message that this producer wanted to hear. He told me quite succinctly that his program would deal with the future of evolution in the *absence* of humanity—that surely humans would soon go extinct. To be fair, I find this view almost universal among my acquaintances as well. There seems to be an underlying human belief that *Homo sapiens* will soon join *Tyrannosaurus rex* and the dodo in the pile of evolutionary discards—basically a guilt and shame response, I surmise ("Anything as bad as we humans will surely die off soon! Why, we may blow ourselves up next week!").

"But what could kill off humanity?" I asked. He responded with the familiar litany: war, disease, asteroid impact, famine, climate change. Besides, he added as an afterthought, the future of animals and plants would be so much more *interesting* without humans—by interesting, I divined that he meant that it would make better television. I asked him to consider my alternative. His answer was that the matter was already decided. The BBC had had a meeting in Bristol, and the decision had been made: humanity would soon go extinct, and the future would be wild—the tentative name of the program. Perhaps a plaque should be mounted somewhere in Bristol inscribed as follows: "On this spot in 1999, executives of the British Broadcasting Company decided the future of the human race—and of all future evolution."

In one regard Caius Julian of the BBC and I are in accord: The future *is* wild, and of this I have no doubt. But, in my opinion, not wild in the way that the BBC might think. It is far more likely that the future will be wild in the way that kayak builder and former tree house dweller George Dyson thinks—a digital wilderness of humans co-evolving with machines, or a wilderness of genetically altered plants escaping from agricultural fields to change the world into a landscape of weeds, or a wilderness of cloned sheep walking amok among their even more staid and normally bred brethren.

There was silence on both ends of the line, and I realized that I had been daydreaming. Finally, it was time for the real pitch: would I consider filling the post of scientific advisor for the series? But we both knew that was a nonstarter now, for my view is that while the future will indeed be wild, many of its evolutionary products will be tame—further domesticated vassals of humanity. The reasoning behind this conclusion, outlined in the ensuing chapters, comes from my lifetime of walking ancient outcrops and visiting gravesites of the geologic past. This book might be far more entertaining if I took the road of the BBC, or of a visionary named Dougal Dixon, and portrayed an interesting bestiary evolving in a new Eden following the fall of humanity. But I do not think that those paths are anything but fantasy.

This book is a look back and a look forward into worlds past and worlds perhaps to come. How can this vision both backward and forward into time be told? Simple prose will do some of the job, but thousand-words-worth pictures will do it as well, or perhaps better. My partner and frequent inspiration in all of this has been artist Alexis Rockman, my equally dark twin. Our methodology was simple: each morning, to the delight of the stockholders of Ma Bell, we spoke on the phone, holding conversations about art, science, basketball, movies, and new visions of what might come next in the history of life on this planet. Words spoken would speciate into words written and pictures painted, followed by faxes sent across the continent or the oceans, depending on the apogees of life's travels. Sometimes we sent each other as well; he to live with me and help me buy books, and I to sleep on his miserable couch, buy expensive New York take-out food, and live in his windowless studio where vision becomes visible; I colored his paintings and he shadowed my moods and thoughts about the future of evolution. I am a scientist but he is a naturalist, and therein lies a harmony often ending in cacophony, for the future may not be pretty, and the past has surely been brutal. So here art and science will collide as well, as Alexis Rockman takes these precepts and mediates them into images. It is our collective vision of a bit of the past and more of the future of evolution. His hands were on these keys, and mine on his brushes.

Others have also had an influence: the evolutionary scientists, of course, notably Norman Myers, Martin Wells, Robert Paine, and Gordon Orians; my colleagues in the extinction business, including my Permian companions Roger Smith and the Karoo paleontology field crew; James Kitching, Joe Kirschvink, the crew from the Foundation for the Future, and especially Sir Crispin Tickell; Dr. David Commings, Neal Stephenson, George Dyson, our agent Sam Fleischmann, and our editor John Michel. Thanks to Holly Hodder for books, and to our families for patience.

Here goes—

FUTURE EVOLUTION

*The farm.*

INTRODUCTION

# THE CHRONIC ARGONAUTS

Nothing in biology makes sense except in the light of evolution.
—THEODOSIUS DOBZHANSKY

Cambridge lies well east and north of London, nestled in a flat landscape softened by time. The spacious farms surrounding this ancient college city are furrowed in white and brown, for the plows gutter into the white chalk making up this part of the British Isles. The chalk comes from a different time; it is a legacy of a long-ago tropical sea filled with the Cretaceous bestiary of a saurian world, an era when dinosaurs ruled and seemingly had all the time in the world to revel in their hegemony. In the oceans the dominant creatures were many-tentacled ammonites, relations of the modern-day octopus and squid. Now they and their world are but memories in chalk, to be disinterred each plowing season.

Slender lanes lead from the center of Cambridge and its splendid University to both rustic working farms and more genteel estates, many of some age. One such manor house sits amid hedges and spacious gardens going wild; around back a large pond long ago given over to eutrophication reflects back the gray skies and slanting rain, while ancient trees afford some slight protection from the English weather for more zealous players on the croquet pitch. The ivy-covered house, stone cold in the grand English tradition, counts its age in centuries. A huge kitchen is its warmth, but the book-lined study is its heart. Like many old English houses, it is a hodgepodge of rooms and uneven floors, the results of successive owners adding on here and there, burrowing amid its warrens, building a wall or tearing one down, marking the centuries with their successive versions of home improvement. Deep in the house's center a great clock ticks, marking time's unidirectional progress, while deeper still the ghost of H. G. Wells just might reside.

The current owners are people of the University. Martin Wells is a professor of zoology; his wife Joyce is a financial officer. Martin has had an influential scientific career, one now far nearer its end than its beginning; he started out investigating octopuses in Naples as part of his graduate studies and continued for years after that, probing the consciousness of these arcane kraakens, puzzling over their eyesight and superb reflexes, wondering how their large brains worked. Later he moved on to explore the brains and physiology of other cephalopods, including the most ancient of all, the chambered nautilus.

It was on an expedition to study *Nautilus* that I first met him. We lived together on an isolated island in the Great Barrier Reef, and sailed the antipodean seas in the sun-drenched tropics to probe the most ancient of living things. I remember thinking then that Martin was slumming a bit in his nautilus studies. His first love remained the octopuses, those creatures that served as models for Martians in the most celebrated book by his grandfather, the English writer and prophet H. G. Wells. *The War of the Worlds* became famous for its malevolent molluscan Martians. Did H. G. ever talk to his grandson Martin about these octopus-like invaders? Somehow, in all of our long days and nights together, I never asked him; perhaps he told me, but time has erased the memory. Perhaps cephalopod preoccupation runs in the Wells family, an odd recessive gene.

H. G. was a Londoner, not of Cambridge, and he never visited the manor house now serving as the ancestral home. But if anything spiritual of H. G. still exists anywhere, it must be in this house. His memorabilia, the numerous first editions, even the remaining royalties from the great man's publishing empire make their way here. This was not H. G.'s house during his lifetime, but it is now.

I first came to this place on a cold March day now many years ago and stayed for a week, playing croquet with Martin, drinking his elder flower wine, and plotting new research on our favorite creatures. Here he critiqued and corrected the draft of my first book, a scientific treatise on the nautilus. We talked endlessly amid the playing and drinking, and when late at night I shivered under the piles of covers in my unheated room and listened to the ticking of clocks, it was of H. G. that I thought, imagining his life, and wondering where his inspiration welled from.

I prowled the house and perused the many books. One of my interesting discoveries concerned H. G. Wells's first book of fiction, published in the final years of the nineteenth century. At that time, Wells was surrounded by the same end-of-century hysteria that deluged my own world as the twentieth century (and second millennium) came to a close, and surely then (as now) all eyes gazed forward toward the uncertain future. Those of H. G. certainly did. His first novel remains among his best known, a rather brief story about a man who builds a machine that can travel through time. Given the choice of voyaging either forward or backward in time, he (like Wells) is interested only in the future. His motive is simple: to see the future of humanity. The name of this novel was *The Chronic Argonauts.* It was later renamed *The Time Machine,* and literary history was made.

Hollywood and armies of pulp science fiction writers have made this story (and this genre) now instantly recognizable to us. The wonderful 1960 George Pal movie of the same name is still staple fare to those surfing across the television bandwidth late at night. But the movie plot departs significantly from the novel it was based upon, and the novel, now not so widely read, holds a number of surprises. The hero, an unnamed Time Traveler, journeys eight hundred and two thousand years into the future from the site of present-day London, and finds wonders—and horrors. Humans have changed—they have undergone evolution. They have become smaller in stature, and more feminine: the men have no facial hair, their mouths and ears have been reduced in size, their chins are small and pointed, their eyes "large and mild." And it is not just the people that have undergone what might be called "future evolution." The Time Traveler finds himself in an ecological habitat very different from the England of today (or of Wells's yesterday). Wells portrays a world where the very plants have changed. Early in the book, Wells's protagonist describes this future world in the following fashion:

> My general impression of the world I saw over their head was of a tangled waste of beautiful bushes and flowers, a long neglected yet weedless garden. I saw a number of tall spikes of strange white flowers, measuring a foot across the spread of their waxen petals. They grew scattered, as if wild, among the variegated shrubs.

But the garden is not so weedless, it turns out, for it is the former food crops that have escaped from the gardens and fields of Wells's time to become the weeds of the future. The Time Traveler also finds that the human inhabitants, the Eloi, are vegetarians. Some of the fruits that they eat are of new varieties. Even the flowers are different—and most animals are now gone. Clearly a vast extinction has occurred, and a great deal of evolution has transpired. All is not novel, though, for Wells has populated this future world with some old standbys familiar in our world, including rhododendrons, apple trees, acacias, tree ferns, and evergreen trees, as well as the new types of fruits and vegetables.

The Time Traveler has seemingly landed in a Garden of Eden. The well-known plot quickly refutes that first impression, however, for Wells has populated his future world with a second human species—the Morlocks, a troglodyte race small in size, of apelike posture, with "strange large grayish-red eyes" and white flaxen hair. Wells is quite clear about the affinity of this group of creatures:

> Gradually, the truth dawned on me: that Man had not remained one species, but had differentiated into two distinct animals: that my graceful children of the Upper World were not the sole descendants of our generation, but that this bleached, obscene, nocturnal Thing, which had flashed before me, was also heir to all the ages.

*The Time Machine* was first published in book form in 1895, and later reprinted innumerable times. Yet prior to its book publication it appeared in serial form in the *National Review,* and that version contained several pages of text omitted from all subsequent book editions. In these pages Wells amplifies his prediction about the fate of animals: by 800,000 years from now, humanity will have killed off almost all of the world's animals, "sparing only a few of the more ornamental." Wells is describing a mass extinction produced by the actions of humanity. There is a clear message in this novel written in the late 1890s: the plants, animals, and humans of the future will evolve from their state in the present, but many of the extant species of our world will not have a future: they will be driven to extinction by humanity.

Late in the book, there is a final, terrible prediction. The Time Traveler voyages many millions of years into the future. The sun has turned orange. Plant and animal life is sparse; he finds giant insects to be the dominant inhabitants of the Earth. The human race still exists, but has "devolved" into small creatures that look like rabbits or kangaroos. It is a dark and depressing chapter in a book already dark and hopeless in tone. The future of humanity is not extinction, it is evolution—but it is not a very "progressive" evolution, at least as many of us would like to define human progress. We do not end up as wiser, more beautiful, more refined creatures. Quite the contrary.

H. G. Wells made a number of unambiguous predictions in *The Time Machine.* First, the book clearly implies that evolution will continue in the future. Second, humanity will create a great mass extinction on Earth. Third, the surviving future flora will be filled with agricultural species run riot and turned into weeds. Finally, humanity itself is virtually extinction-proof, though it will evolve. Wells was, of course, a confirmed evolutionist. He passed his college years at the Normal School of Science in London, where he took classes on evolution from Thomas H. Huxley himself. *The Time Machine* is a science fiction novel, one of the first ever, but above all it is an early and prescient attempt to chart the future of evolution. A century later it is difficult not to concur with its predictions.

## The Future of Evolution

What is the future of evolution? So ambiguous a question invites varied responses. As in *The Time Machine,* it might be interpreted in terms of outcomes: what will animals, plants, and other organisms be like at some time in the future, perhaps a thousand years from now, perhaps a thousand million years from now? The only certainty is that they will be different. Even in the near future, the mix of species and their distributions, relative numbers, and relationships with one another will have changed, and by the far future the accumulated changes may be breathtaking—or trivial. There can be no doubt that the evolutionary forces that have created the astonishing diversity of species on Earth in the past and into the present will continue creating new species and varieties, resulting in a global biotic inventory of species different from that of today. How different, and in what ways, is open to informed speculation, and is one of the subjects of this book. This particular question was addressed some years ago by author Dougal Dixon in his delightful 1970 book *After Man.*

Ahead of his time (if still well after H. G. Wells), Dixon echoed Wells in forecasting an imminent mass extinction, prophesying that humanity would eliminate enough of the current biota on Earth to open the faucets of evolutionary change. But here Dixon parted company with the Wells vision, for Dixon posited his new fauna evolving in a world where humanity itself has gone extinct. Dixon predicted that most of Earth's post-extinction bestiary would evolve from the surviving meek, such as small birds, amphibians, rodents, and rabbits. Dixon's central assumption is that humanity will biotically impoverish the planet and then have the good grace to go extinct, opening the way for the evolution of many new species. His imagined new biota depends on this central fact—that humans have gone extinct, yet left the Earth in sufficiently good repair to allow wholesale evolution of new forms. The creatures figured show *evolutionary convergence:* they resemble the animals that *might* soon be extinct on the present-day Earth. Dixon has thus figured animals resembling the many endangered large herbivores, carnivores, and scavengers in the varied biomes represented on the planet today. While great fun, this new fauna (like that portrayed in far less detail in *The Time Machine*) is a completely untestable vision residing in the realm of fantasy.

The pathway of Dougal Dixon—of imagining a subsequent fauna and flora—is one way of answering the query about the future of evolution on Earth. Yet there is another way that the question might be interpreted. Perhaps it relates not to *outcome,* but to the evolutionary *process* itself. It might mean, "What is in store for the varied mechanisms that have resulted in the various species of the past and present?" Might the "rules" governing those processes be changed in the near future—or might they have been changed already in the not so distant past?

The second interpretation of this question seems, at first, patently ridiculous. The processes that introduce novelty and evolutionary change—natural selection, mutation, sexual selection—will continue to modify the gene pools of species, occasionally resulting in the formation of new species, just as they have since life first appeared on the planet at least 3.8 billion years ago. But it may be that while *process* has not changed, *pattern* has. This is the point of another English thinker, Dr. Norman Myers of Oxford University. One of the most vocal and prominent conservationists of the late twentieth century, Myers believes that humanity has changed the rules of speciation itself. That controversial view will also be explored in the pages that follow.

Whether or not we have somehow changed fundamental aspects of how or where new species arise, it is an unambiguous fact that very early on, our species learned to manipulate the forces of evolution to suit its own purposes, creating varieties of animals and plants that would never have appeared on Earth in the absence of our will. Large-scale bioengineering was under way well before the invention of written language. We call this process domestication, but it was nothing less than efficient and ruthless bioengineering of food stocks—and the elimination of species posing a threat to those food stocks. Once the new breeds of domestic animals and plants became necessary for our species' survival, wholesale efforts toward the eradication of the predators of these new and stupid animals were undertaken. A carnivore eating humans was tolerable, because the losses were negligible, but a carnivore eating the new human food sources was not, because the losses spread to the entire group.

Our modern efforts at biological engineering are but an extension of our earlier efforts at "domestication." Until the end of the twentieth century the natural world had never evolved a square tomato, or any of the numerous other genetically altered plants and even animals now quite common in agricultural fields and scientific laboratories. Just as physicists are bringing *un*natural elements into existence in the natural world through technological processes, so too has our species invented new ways of bringing forth varieties of plants and animals that would never have graced the planet but for the hand of man. And, like plutonium, the new genes created and spliced into existing organisms to create new varieties of life will have a very long half-life; some may exist until life is ultimately snuffed out by an expanding sun some billions of years in the future. So what is the future of evolution? Some of it is being decided in biotechnology labs at this moment.

Humans have profoundly altered the biotic makeup of the Earth. We have done it in ways both subtle and blunt. We have set fire to entire continents, resulting in the presence of fire-resistant plants in landscapes where such species existed only in small numbers prior to the arrival or evolution of brand-bearing humans. We have wiped out entire species and decimated countless more, either to suit our needs for food or security or simply as an accidental by-product of our changing the landscape to favor our new agricultural endeavors. We have changed the role of natural selection by favoring some species that could never otherwise survive in a cruel Darwinian world over others of estimably greater fitness. We have created new types of organisms, first with animal and plant husbandry and later with sophisticated manipulation and splicing of the genetic codes of various organisms of interest to us. The presence of humanity began a radical revision of the diversity of life on Earth—both in the number of species present and in their abundance relative to one another.

It is not just modern humans in gleaming laboratories that have instituted this biotic change, or even the primitive farmers that caused the evolution of the now familiar domesticated animals beginning 10,000 years ago. Hunters have also significantly participated in creating evolutionary change that will echo through time for thousands or tens of thousands of years still to come. We have not only created new ways of producing animals and plants through brutal *un*natural selection, but we have manipulated the most potent force of evolutionary change—the phenomenon of mass extinction. Humanity has created a new mass extinction—which I will show to be now largely over—that is different from any that has ever affected the planet.

A central thesis of this book is that the most consequential aspects of the new mass extinction of species so direly predicted to be awaiting us in the wings of the near future have, in fact, already occurred, at least among those creatures that contribute most importantly to the makeup of the terrestrial biosphere. One of the tenets of the modern evolutionary theory known as *macroevolution* is that mass extinction is a potent source of new species once the forces that brought it on have ebbed. According to this theory, the elimination of a majority of species (characteristic of the direst mass extinction events) opens up opportunities for new evolutionary varieties to fill the rolls of the missing. The end result is that novelty reappears on the planet. Many evolutionists have theorized that over the next several centuries humanity will directly or indirectly create just such a situation. In contrast to this position, I argue that, at least for the most important terrestrial animals, this has already happened.

The elimination of large mammals during the last 50,000 years has profoundly affected the evolutionary mix of the planet, and should create the opportunity for a new evolutionary fauna to arise—much like new plant growth following a forest fire, but in this case composed of entirely new types of species. Just such recoveries followed the two greatest mass extinctions of the past: the Permo-Triassic extinction 250 million years ago, which ended the Paleozoic Era of life and ushered in the Mesozoic, and the Cretaceous-Tertiary extinction 65 million years ago, which ended the Mesozoic and created the conditions leading to the Cenozoic Era.

The first of these mass extinctions caused a changeover from a terrestrial world dominated by mammal-like reptiles to one dominated by dinosaurs, while the second opened the way for the Age of Mammals with the complete extinction of the dinosaurs. These and other catastrophic mass extinctions in the Earth's past were invariably followed by periods when the Earth was inhabited by a relatively low number of species, known as recovery faunas. These depauperate recovery faunas were in turn succeeded by a newly evolved group of dominant organisms, often composed of taxa different from those that dominated prior to the mass extinction.

So too with the extinction of the Ice Age megamammals, which I see as simply the opening (yet most consequential) act of a mass extinction continuing into the present day. This "modern" mass extinction has been profiled in a slew of recent books and articles, such as Ehrlich and Ehrlich's *Extinction,* my own *The End of Evolution,* Niles Eldredge's *The Miner's Canary,* Leakey and Lewin's *The Sixth Extinction,* and David Quammen's *The Song of the Dodo.* If the past is a key to the present and future, we can expect the emergence of some new Age—an age of new varieties of mammals, or an Age of Birds, or perhaps an Age of animals of a body plan yet to be evolved.

Or maybe not. There are some doomsayers who suppose that there will *not* be good news after the bad news, or at least any time soon after the bad news. According to this school of thought, there will indeed be a new age: an Age of Weeds, or perhaps a Depauperate Age. The most prominent of these thinkers is Norman Myers, who brings up an intriguing and disturbing point: What if the processes and places that have restocked the biodiversity cupboard in the past can no longer operate because of the way that humankind has reshaped the surface of this planet? In the past the tropics have repeatedly reseeded the Earth with species. But it is these same tropical regions, and especially the tropical rainforests, that are being most radically affected by burgeoning human populations. Thus Myers believes that there will not be much of a new recovery fauna for many millions of years into the future—if ever.

Yet there is a third alternative. What if we are *already* in the midst of the new Age? What if the dominant organisms of the new evolutionary recovery fauna have *already* evolved? In the pages below I will try to show that a newly evolved recovery fauna is indeed already among us, composed of new types of mammals and birds unknown on Earth even 15,000 years ago: cows, sheep, pigs, dogs, cats, chickens, pigeons, barnyard ducks and geese, and fluffy Easter bunnies, among others: the familiar domesticated animals that serve as our companions and food sources. This is not to say that there will be no further extinctions in the centuries to come as humanity continues to enlarge its numbers and its hold on the Earth. But the subsequent species losses will have very little further evolutionary effect on the planet, and will occur mainly among terrestrial species of little evolutionary importance—species that will probably not be replaced as long as humanity exists.

My final proposition is that of all large animals (and certainly among large mammals) on Earth, save for the bacteria inhabiting the deep microbial biosphere of the Earth's upper stony crust, our species is the most extinction-proof, unless a very low probability traumatic event, such as a very large asteroid or comet impact or an all-out nuclear war, comes along. Yet even in the latter case there is still a high probability that some few of our resourceful species will emerge from some bomb shelter and return to our rabbity breeding ways. This is not to say that our species will be happy, but exist we will—and as long as that happens, there will not be a new age of anything except a continued Age of Humanity.

Even in such a world there will be future evolution, and there will be new varieties of animals and plants. The current use of plants and animals with altered genomes—transgenics—ensures that the future will look different from the present, especially if one appreciates the beauty of ragweed, or can appreciate pesticide-resistant horseflies. The future may be one of runaway giant pumpkin vines and other escaped, altered agricultural crops. It will certainly be interesting.

This book is not about cause. It is about effect. I will describe in some detail the major mass extinctions that have already occurred, but while most treatments of mass extinction focus on causes, that is irrelevant to the arguments in this book—it hardly matters whether an extinction was brought about by climate change, meteor impact, or human activity. All achieve the same effect—our story.

Eight propositions to be defended, then:

1. Past mass extinctions have been instigators of biological innovation and the eventual augmentation of diversity. They have opened up ecological niches and fostered the creation of evolutionary novelty.
2. Most—or all—past mass extinctions have been multi-causal, and have lasted tens of thousands of years at a minimum.
3. The Earth entered a new mass extinction event during the waning of the last Ice Age—a mass extinction that continues into the present.

It is likely to continue well into the future. But its most consequential phase—the destruction of large mammals and birds—is already finished, and it happened (at least by human standards of time) a very long time ago. It resulted in the extinction of the dominant terrestrial organisms, the large *megamammals* that populated most of the land surface until the last phases of the Ice Age and into the present. This new mass extinction now preys upon the small, the endemic, and the wild species such as salmon and cod that are harvested as human food. But mostly it preys on animals and plants living on biotic islands—either real islands surrounded by water or the artificially produced habitat islands that our highways and croplands are creating. What we witness now is a highly significant yet almost invisible diminution of the smaller species on Earth, for the larger animals are already gone.

4. The modern mass extinction is different from any other in the Earth's long history.

To date, it has affected mainly large land animals, island birds, and rare tropical species, although data emerging in recent decades suggest that its highest extinction rates may be shifting to tropical plant communities and perhaps tropical marine coral reefs. It is certainly causing the depletion of wild food stocks of land and marine animals. The reduction of fishery stocks is causing a wholesale elimination of major populations that may not kill off entire species (due to fish farming), but will leave the planet biotically impoverished nevertheless. Global terrestrial biodiversity will fall to end-Paleozoic levels because of continued extinction and the functional removal of traditional barriers to migration.

5. All mass extinctions have been followed by a recovery interval, characterized by a new fauna composed of animals that have either survived the extinction or evolved from such survivors.

In this case, that recovery fauna is already largely in place, and consists mainly of domesticated animals and plants, as well as "weedy" species capable of living amid high populations of humans.

6. There will be new species yet to evolve.

Many of these new species will be the result of jumping genes, as DNA from organisms created under laboratory conditions by biotechnology firms escapes into the wild. Others will be mainly small species adapted to living in the new world of spreading cities and farms. The new animal and plant species will thus evolve in the niches and corners of a world dominated by *Homo sapiens.* The rules of speciation have changed: few large animals will evolve as long as humanity exists in large numbers, and as long as our planet remains dividend into innumerable small islands.

7. Our species, *Homo sapiens,* can look foreword to both evolution and long-term survival. Of all the animal species on Earth, we may be the *least* susceptible to extinction: humanity is functionally extinction-proof.

Yet we are also malleable by the evolutionary forces of natural selection, and we may be seeing rapid evolution within our species at the present time, as evidenced by an increase in the incidence of potentially heritable behavioral disorders (attention deficit hyperactivity disorder, Tourette's syndrome, clinical depression). There will also be what might be called "unnatural selection" as some segments of humanity acquire the use of neural connections to sophisticated memory storage devices. The future evolution of humanity will entail integration with machines—or perhaps we are but the midwives of the next global intelligence: machine intelligence.

8. There will never be a new dominant fauna on Earth other than humanity and its domesticated vassals until we go extinct—and if we succeed in reaching the stars, that may never happen.

Prophecy is perilous business. But there are some clues, mainly from the fossil record, about how the future of evolution may proceed. These clues and their implications are the subject of this book.

*The mass dying out of the late Permian Period was the greatest of all extinctions. Although some animals, like this burrowing Lystrosaurus, may have tried to escape their fate, eventually 90% of all the animals on Earth disappeared. This scene will repeat itself at the end of the world.*

ONE

# THE DEEP PAST
## A Tale of Two Extinctions

Force maketh nature more violent in the Returne.
—FRANCIS BACON

### Karoo Desert, South Africa

On a cloudless September day a paleontologist gears up for a collecting trip in the dry Karoo Desert of southern Africa. His route will cover a distance of about 3 kilometers, beginning at the base of a large valley overlooked by a high mountain ridge known as Lootsberg Pass. The traverse will take him through time as well as space. As he climbs up through the bed of an ephemeral creek he will be ascending a stairway made of stacked layers of sedimentary rock, each stratum representing a slice of time, starting with 251-million-year-old strata and ending in 249-million-year-old rocks. Somewhere on this walk he will pass through the remains of a singular biodiversity catastrophe, the single most calamitous mass extinction to have ever savaged the Earth, an event so severe that it has forced geologists to subdivide time around it. He begins his traverse in Permian rocks, representing the last time interval of what is termed the Paleozoic Era, so named because of its archaic assemblage of fossils. He will end up in the Triassic period, the first unit of the Mesozoic Era, or "time of middle life." The division between these two groups of strata was caused by mass extinction. The various tools of his trade are attached to or slung on the hooks, holsters, belts, and vest he wears; water and food are doled out and stored in the backpack that completes his burden. A broad-brimmed hat tops it all off, and he laughs at himself—Halloween in Africa, a geologist in costume. With no more ceremony than locking the vehicle doors, he sets out into a basin tens of kilometers across.

His first impression is of heat, first felt against his rapidly drying skin, then glimpsed as faintly perceived shimmers in the clear air. Great vultures ride these thermals, but otherwise the vista is exanimate. Only a thin ribbon of road brings order to the wide valley floor, and a sense that the living share this place with the fossil dead. The air is so clear that great distances lie visible, as if the landscape is part of another, larger planet, where the very horizon recedes impossibly far, or perhaps this world of dreams is flat. Green fights a losing battle among the low, dispirited shrubs and thorny scrub, fading to relentless brown, a thousand shades of brown; a color elsewhere so monotonous yet here so diversified.

From his vantage point at the base of the valley, the world seems to have encountered a great test, and been found wanting. A test of life failed; failed in the present, failed immeasurably more so in the quarter-billion-year-old past, for Lootsberg Pass is a fossil graveyard, a headstone to Planet Earth's greatest extinction.

Three hundred million years ago, in a time long before dinosaurs, mammals, or birds had first evolved, the southern part of what we now call Africa was gripped in the glacial deep freeze of a profound ice age. Slowly, the land warmed, and a landscape suitable for life emerged. First low mosses, then higher forms of life colonized the rapidly warming region, eventually creating a lush world of wide river valleys far from the sea. Into this region animals found their way, and thrived. They left their remains in the ancient river sediments, remains that only now are eroding free in the isolated sedimentary rock banks and outcrops beneath Lootsberg Pass.

The geologist makes his way to low outcrops of greenish sedimentary rock carved into the grass and scrubland making up the wide valley floor. The sedimentary rock beds in this or any other exposure are windows to the deep past, for it is within such strata that information about ancient environments, as well as ancient inhabitants, is entombed. Because of their textures and bed form, these particular sedimentary rocks could have formed only in rivers. The rocks also bear *fossils,* remains of ancient plants and animals.

The river valleys of 250 million years ago would have looked much like any river valley today, with meandering streams and swamps. But the rich plant life would probably seem exotic and peculiar to us if we could somehow be transported back to this ancient time. While the world today is dominated by flowering plants, the fossils in these greenish river deposits are from species far more ancient: mosses, ferns, club mosses, ancient horsetails, and most commonly, seed ferns of a type called

*Glossopteris* (the modern-day ginkgo is a descendant, and gives us a sense of what this plant may have looked like). Giant horsetails in stands like bamboo might have lined the riverbanks. Ferns, mosses, and primitive plants known as lycopods were also common. There might have been savanna-like regions as well (but without grass, a much later innovation). Paleobotanist Bruce Tiffany envisions the Karoo vegetation as "gallery forests," isolated stands and thickets composed of seed ferns, with conifer trees in areas of moisture, surrounded by regions of true ferns. The ferns may have formed extensive communities, almost like grasslands. All of this richness fringed the watercourses; in more upland regions, away from water, there may have been little vegetation. All in all, it was an ideal place for land life.

At first only squat, belly-dragging amphibians lived in these river valleys. But as eons passed, more advanced land dwellers arrived or evolved: the fully terrestrial reptiles, small at first but rapidly enlarging until a great diversity of spectacular and hulking behemoths waddled and shuffled about the landscape. Several stocks lived in this ancient African splendor. Most common were four-legged creatures called therapsids, or "mammal-like reptiles." But other reptilian legions spawned here as well, such as the ancestors of turtles, crocodiles, lizards, and, eventually, the dinosaurs. Some were hunters, far more were hunted. All have left a copious fossil record of their presence, for the Karoo strata are packed with bones.

The therapsids are virtually unknown to us in any sort of cultural context; theirs is the true lost world. When in Edwardian times Sir Arthur Conan Doyle wrote his scientific adventure story *The Lost World,* he recreated an environment known at that time only to academics: the world of the Mesozoic Era, known to us as the Age of Dinosaurs. He created a place lost in the world because of geographic isolation, but he was really painting a picture of scientific isolation, for even in the early twentieth century the great Age of Dinosaurs was still a lost world, so little did science (and the public) know about it. The Age of Dinosaurs is clearly no longer so lost. Every schoolchild knows the dinosaurs' tongue-twisting names, their food preferences, and even their color schemes. Nothing so well known to Hollywood and popular culture can be considered lost. Instead, the true lost world is that of the mammal-like reptiles—a time and place that disappeared from the Earth a quarter billion years ago.

Paleontologists now have a fairly accurate census of the large vertebrate genera living in the Karoo Basin just prior to the great extinction. There were two amphibian genera (and thus at least two, but probably more, species), six types of captorhinids (ancestors of turtles), two eosuchians (ancestors of dinosaurs, crocodiles, and birds),

*The* T. rex *of its time, the gorgon was the largest of the Paleozoic predators. The drawings here represent four possible renditions of what this animal might have looked like.*

nine mammal-like reptiles known as dicynodonts (which shared a common ancestor with mammals), three biarmosuchians (a primitive group of reptiles), nine gorgonopsians (all large, fearsome predators), ten therocephalians (another group of now-extinct reptiles), and three cynodonts—doglike predators that are on the direct line of all living mammals. All told, forty-five separate genera of vertebrate creatures are known from this last million years of the Permian Period.

This census demonstrates that life was diverse in the Age *before* Dinosaurs. To put this number in context, there were fewer large vertebrate genera in the Permian Period than, say, on the plains of modern-day Africa or in the rainforests of the present day. But there were *more* large animals back then than are found today in the grassy regions of North America, or Australia, or Europe, or Asia. This ancient world was diverse, in some cases more diverse than our own in the category of large, four-legged land life.

Until the highest reaches of Permian-aged rocks, there appears to be no diminution of either the numbers or diversity of the Permian fauna as one approaches the boundary marking the mass extinction. The most common fossil is *Dicynodon,* name-giver of this highest Permian zone, but many other types are found as well. As in any environment of today, the herbivorous forms far outnumber the predators. Then, very curious changes begin to appear in the rock record.

About halfway up the gully fronting the Lootsberg Pass region, the rocks begin to change color from greenish to red. The green and olive strata first show faint patches of purple, and as successive strata are passed on a journey up through the great stratal column making up this region, more and more of these red to purplish blotches are found within the rocks. Another change occurs as well: the fossils become more rare and far less diverse. Forty feet above the first appearance of reddish strata, only three types of fossils can be found, and two of these were not present in the greenish strata below. *Dicynodon* is still present, but it is now the only member of the impressive diversity of Permian fauna that was so commonly found in the strata below. The two new fossil types that appear are a small but vicious-looking predator called *Moschorhinus* and a curious dicynodont genus called *Lystrosaurus.* Elsewhere in the Karoo, a few other types are known from this interval as well, including a small lizardlike form, some amphibians, creatures looking something like a dog, and a small reptile that turns out to be the ancestor of the dinosaurs.

*Dicynodon, Moschorhinus,* and *Lystrosaurus* are found together in beds over a stratal thickness of perhaps 50 feet or so. For the last 10 feet of this interval the beds are pure red; they have lost any semblance of green color. And then a most curious sedimentary phenomenon occurs: one last time, green beds appear. The most distinctive of these beds are found in the Lootsberg gully—and, it turns out, everywhere else in the Karoo at this same stratigraphic interval so far examined. These last green beds are very thinly laminated, showing the finest-scale bedding planes and sedimentary structures. They have no burrows, no evidence of plant material—and no fossils of vertebrate animals. They are completely barren, with an aggregate thickness of only 10 feet. They are the signposts of global catastrophe.

The strata immediately above and below these thin, green laminated beds show no bedding planes and are red in color. The lack of distinct bedding in the underlying and overlying strata comes from a process known as bioturbation, caused by the action of burrowing organisms such as insects, worms, and crustaceans, which disrupts the original bedding, making it gradually indistinct. Almost all sedimentary beds are thinly laminated when they are first deposited. But in most environments in our time (and probably in most of the Permian time as well), the action of burrowing animals disrupts this fine-scale bedding. As years and then centuries pass, the fine-scale differences in sediment composition producing the visible bedding planes are destroyed, trampled, ingested, homogenized. The resulting rock is massive, featureless, and free of bedding plane surfaces. Oddly enough, it is the presence of fine bedding planes that alerts the geologist to the fact that something extraordinary has happened, for the presence of such beds indicates that organisms were not present. It tells of a world existing in the absence or near absence of animals. And that is rare indeed.

The sun rises higher in the clear sky; the geologist is halfway through his trek. The heat of the day bares its fangs; sweat emerges on his skin only to dry instantly in the hot wind. He feels like an inverted diver; he drinks from the large water bottles he carries, filling himself with water like some lost fish emerging onto land in a diving suit that pumps water, rather than air, into his body. The vegetation around him is all scrubby, low, and brown; an occasional carnivorous fly buzzes about his face, intrigued by this moving, sweaty heat source. More water, salted nuts, an orange, an apple, and he shrugs on the heavy pack once more and continues upward.

The rocks are very different now. All of the finer rocks are brick red in color. It is like the surface of Mars—perhaps in more ways than one. The geologist comes to a thick ledge of sandstone, and finds pebbles and bones along the undersurfaces of these thick beds. They show features indicating that they were deposited by braided streams, the anastomosing channels that water follows as it first leaves the mountains, or on any other steep slope. There is no evidence of the more meandering

*Inheritors of the post-Permian world, the dinosaurs would quickly dominate in species and individuals.*

rivers of a typical river valley, no indication of the cross-beds and point-bar deposits that all rivers past and present make when they cross a river valley. Such deposits are common in the green beds of the Permian rocks seen earlier on this trek, but have disappeared from the Triassic strata. He wonders, imagines the scene. Perhaps the land suddenly tilted upward, creating a slope where none was before; mountain building could do that. But there is no other evidence that the long-ago region of Lootsberg Pass was affected by rapid mountain building.

He searches his long memory, and Mars and H. G. Wells come to mind. Long ago there was water on Mars, and rivers. But all of the rivers on Mars were braided, leaving behind the same types of deposits, he is sure, that are found in these lowest Triassic strata in the Karoo. The reason the rivers on Mars were braided is that there was nothing to stabilize their banks, no deep roots to hold them in check, for evolution there, if it produced life at all, probably never got beyond bacteria. And the connection clicks. He has a vision of a long-ago Earth, where rivers were always braided—until plant life evolved and introduced a new type of river, the meandering river so familiar to us all in our world, and familiar too in the Permian period. Then, 250 million years ago, a huge mass extinction made this portion of the Earth, and perhaps all of the Earth, suddenly Mars-like, stripped it of all of the Permian trees and bushes that had greened that ancient world and kept its rivers flowing in the sinuous and meandering channels so recognizable and familiar to those of us who live in a tree-filled age. It hits him: this ancient extinction killed off the Permian trees, perhaps most of the Permian plants. And in so doing it changed the way the rivers flowed.

The day is nearly over when he finishes his climb. In the highest rocks he finds numerous fossils, mostly of the pig-sized *Lystrosaurus.* But he sees other fossils as well, one that will give rise to the mammals, and another that will be the seed stock of an entirely different group: the dinosaurs, those heirs to the Paleozoic world whose own world also ended abruptly in global catastrophe and mass extinction, in an event best studied along a scenic seacoast in France.

## Hendaye, France

Long ago, Spain whirled in its continental drifting, made a hard right turn, and ran into France with a tectonic lunge. Rocks crumbled, and the Pyrenees became the zipper uniting these two great blocks. An ancient seafloor was raised in the process.

Today a part of that ancient ocean is exposed for all to see, but like Gomorrah and Sodom, that deep-sea bottom and its trove of skeletons has been turned to stone. Now it is a scenic park on the border between Spain and France, a coastal bit of the Basque country. On a very hot day a geologist prepares to hike this bit of coast in order to visit one of the world's most impressive Cretaceous-Tertiary boundary sites, a place where the great catastrophe ending the Age of Dinosaurs is preserved in dramatic fashion. To get there, he has to walk a pathway only the twentieth century could have built, a pathway containing clues not only to the past, but to the future—the future of evolution—as well.

He starts his trek along a busy scenic coastal road lined with "snacks" and open-air cafes, then strides onto a wide sandy beach covered with naked humans. A lone, futile sign proclaims *Nudism Interdit!* (Nudism Forbidden). It is July, a hot morning, and already throngs from nearby Spain are jostling with the German tourists for the best bits of littoral territory as they lather their naked bodies with sunscreen amid piles of discarded clothing. Every age and form of humanity spreads itself out to fry in the sun, and the geologist is an odd sight as he walks through the sand, at times stepping over and by the prone naked bodies, festooned as he is with the hammers, compasses, water bottles, packs, and other regalia of his trade. It is an odd sight to see a clothed man, let alone an *equipped* clothed man. Odder still, he is walking to work, while the rest of humanity is here to frolic in the waves, playing the odd game of Spanish paddleball. The tide of humanity washed onto this shore is oblivious to another flood of flotsam floating in from Spain with the tide: the flotillas of garbage caressing their legs and ankles in the warm Bay of Biscay as they unconsciously celebrate their dominion over a thoroughly tamed world. Not a single one of them worries about being eaten by some predator that day. It is a sure sign that a great mass extinction has taken place: only during mass extinctions do the predators disappear.

At the end of the beach a great rocky headland exposes pinstripes of strata, the Upper Cretaceous sedimentary beds he has come to sample. But the rocks rise precipitously and vertically up from the sea, leaving no path for the beachcomber to pass, so he must climb up onto the headland to get to his target site, still a half mile down the coastline. A well-worn path beckons upward near the end of the beach, and he follows it amid the sweet smell of beach and salt air. The neatly groomed track winds through bracken, then brings him next past a large fenced enclosure filled with children. As he passes closer, he sees that in contrast to the frolic and play normally associated with the young, these children are listless, slow-moving, or motionless. Some are wheeled by white-coated attendants. He realizes that this large outdoor reserve is for autistic and retarded children, all helpless and heart-wrenching in their plight. He walks slowly by, staring, but they take absolutely no notice of him. France has put its most pitiable next to the sea, in an exquisite setting—these children that in another age would die early, but here will live and in many cases breed, and in some cases perpetuate their disabilities. Natural selection is no longer at work for these, or any other, humans.

He ponders this experiment in future evolution as he finally passes by the manicured lawn, itself some new evolutionary joke of grass bred for looks, and the path begins to rise. Now a different assault on his senses occurs: the cool, sweet salt air is suddenly replaced by a gut-wrenching odor, a choking miasma. The path now runs next to the municipality of Hendaye's sewage treatment plant, its huge outdoor pools of cess slowly rotating in giant concrete cisterns. Unfortunately, there is no way over the headland except by this path. A littoral territory once the home of a small tribe of humans is now inhabited by tens of thousands of humans and visited each year by that many and more, and their combined fecal output is now so voluminous that it can no longer be simply dumped into the sea. So here it is "treated" and *then* dumped into the sea, creating a riotous explosion of algal growth in the shallow water around the sewage outfall pipes, an experiment in ecology that is utterly changing the intertidal and subtidal communities along the coast as now bountiful phosphates and nitrates amid their rich liquid fertilizers putrefy the region.

Finally he is past this hurdle as well, and he enters a fairyland. High above the beach a great pasture unfolds: acres of manicured grounds, scattered trees, and the magnificent vista of the sea. Sitting above it all is a splendid spired castle, now housing French astronomers by all accounts, although no telescope can be glimpsed. He is now in the reserve called Abbadia, a huge park that was once the fertile fields of the adjoining castle, and he feels transported back to earlier centuries, with even more distant time travels just ahead. He shoulders through a herd of sheep—animals stupid and bizarre compared with their ancestors. Their fecal pellets lie everywhere, and he wonders if they are, after humans, the most common large mammals on the planet. He ponders the process called *domestication* and how all domesticated animals seem to have lost brainpower as they were sculpted by humans into the species that they have become. He imagines the world of 8,000 years ago as humanity began to populate it with entirely new types of animals and plants in the single greatest evolutionary experiment since the ancient mass extinctions.

As he walks across the high meadow in the sparkling summer sun, the twentieth century and its history once again intrudes. Amid the waving grass, grazing sheep, and linear hedgerows are the scattered remains of huge concrete bunkers, jumbled masses of fractured concrete and twisted rebar. The blockhouses were the work of the Nazis, part of the Atlantic Wall they built for defense, now nothing but large ruins of concrete littering the flat grounds like yawning caves or the litter of capricious giants. A movement within the first broken blockhouse he passes startles him; he expects a fox or dog, but a naked man slowly stands, watching him. He passes by, and another man can be seen in the next smashed bunker. Soon he realizes that the field is alive with half-seen men, all silent, many only partially clothed or, like the first, not wearing clothes at all. He understands suddenly that this park is the territory and cruising ground of the local gay community, a meeting place where the vacationers who come here, and the locals who live here, swap microbes and homogenize the world's infectious diseases. It is a microcosm of what is happening to the world's animals and plants. He wonders, how much of their behavior is genetic, and will that be a future of evolution?

He tops the crest of the headland and begins to drop down toward the sea. A steep and switchbacked trail makes a precarious path to the water's edge, where gently tilted strata are now exposed by the low tide. He strides out onto these rocks, inch- to foot-thick limestone layers packed with the most spectacular fossils.

Giant clams lie frozen in the strata. Not the giant clams of our age that are now seen as birdbaths in backyard gardens, but flatter clams, with huge oval shells as much as a yard in length. They are nothing like any clam now alive, yet once these fossils were dominant members of the Mesozoic sea bottom community. They are called inoceramids, and they are hallmarks of a time when dinosaurs ruled the land and ammonites swam the seas. These same ammonites, with coiled shells like that of the nautilus, are also found in the clam-rich strata, although they are never so numerous as the clams. The geologist notes a few, and begins to walk perpendicular to the bedding, and thus up through time.

It is a beautiful walk, with high cliffs of white limestone and reddish marl arching overhead, the sea slapping the rocks, and gulls wheeling about in noisy cacophony; no clouds mar the deep blue sky. When he has walked along the coast for about 40 meters, the most peculiar thing happens: the clam fossils begin to disappear. Soon they are rarely seen, and then they are gone altogether. They and their kind disappear not only from the strata on this seacoast but from all rocks dated at 67 million years old and less, in which they had been common. After a reign of over 170 million years, this type of clam suddenly goes extinct. The strata look the same, but the giant clams are gone.

The geologist continues his walk along the seacoast, moving relentlessly up through time as he crosses the tilted strata on the rocky beach. Fossils are still present, but they are relatively few in number. Most are sea urchins, although a few small clams and the rare but beautiful ammonites can be seen on this infrequently visited stretch of coast. He passes into a small bay, and the scenery changes. The tan to olive limestone he has been passing over is superseded by a gigantic wall of bright pink rock. There is a clear point of contact between the olive rock and this thicker, pinker limestone, and he moves into the bay to this contact. It is his goal this day. A thin clay layer several inches thick marks the boundary between the olive Cretaceous rocks and the pink rocks of the overlying Tertiary Period. This layer is also where the last ammonites can be found, while its counterpart on land is the stratum with the last dinosaur fossils. He smashes out a few fragments of this claystone with his rock hammer and examines them with a powerful loupe. The clay contains a thin, rusty layer, and under magnification he can see that this thin layer is packed with tiny spherules, invisible to the naked eye but clearly visible even under the low magnification of the loupe. He is looking at bits and pieces of Mexico, on an extended European holiday after being blasted into space by the great asteroid impact that ended the Mesozoic Era 65 million years ago. In the warm sun, on this perfect day, he stretches out on the rocks, one hand on the last of the Cretaceous, the other slightly above it, on the oldest rocks of the Tertiary, spanning two eras, and imagines the scene:

> The asteroid (or comet—who knows!) is perhaps 10 kilometers in diameter, and it enters the Earth's atmosphere traveling at a rate of about 25,000 miles an hour. Yet even at such great speed it can be visually followed as it traces its majestic path down through the atmosphere before finally smashing into the Earth's crust. It is so large that it takes a second for its body to crumble into the Earth. Upon impact, its energy is converted into heat, creating a non-nuclear explosion at least 10,000 times as strong as the blast that would result from mankind's total nuclear arsenal detonating simultaneously. The asteroid hits the equatorial region in the shallow sea then covering the Yucatán, creating a crater as large as the state of New Hampshire. Thousands of tons of rock from ground zero, as well as the entire mass of the asteroid itself, are blasted upward, creating a bar of white light extending up from the Earth into space. Some of this debris goes into Earth orbit, while the heavier material reenters the atmosphere after a suborbital flight and streaks back to Earth as a barrage of meteors. Soon the skies over the entire Earth begin to glow dull brick red from these flashing small meteors. Millions of them fall back to Earth as blazing fireballs, and in the process they ignite the rich, verdant Late Cretaceous

*The Age of Dinosaurs ended when an asteroid crashed into the Earth at Chicxulub on Mexico's Yucatan Peninsula.*

forests; over half the Earth's vegetation burns in the weeks following the impact. A giant fireball also expands upward and laterally from the impact site, carrying with it additional rock material, which obscures the sky as fine dust is transported globally by stratospheric winds. This enormous quantity of rock and dust begins sifting back to Earth over a period of days to weeks. Great dust plumes and billowing smoke from burning forests also rise into the atmosphere, soon creating an Earth-covering pall of darkness.

The impact creates great heat, both on land and in the atmosphere. The shock heating of the atmosphere is sufficient to cause atmospheric oxygen and nitrogen to combine into gaseous nitrous oxide; this gas then changes to nitric acid when combined with rain. The most prodigious and concentrated acid rain in the history of the Earth begins to fall on land and sea, and continues until the upper 300 feet of the world's oceans are sufficiently acid to dissolve calcareous shell material. The impact also creates shock waves spreading outward through the rock from the hole the asteroid punches in the Earth's crust; the Earth is rung like a bell, and earthquakes of unprecedented magnitude occur. Huge tidal waves spread outward from the impact site, eventually washing ashore along the continental shorelines of North America, and perhaps Europe and Africa as well, leaving a trail of destruction in their wake and a monstrous strandline of beached and bloated dinosaur carcasses skewered on uprooted trees. The surviving scavengers of the world are in paradise. The smell of decay is everywhere.

For several months following this fearsome day, no sunlight reaches the Earth's surface. After the initial rise in temperature from the blast itself, the ensuing darkness that settles in causes temperatures to drop precipitously over much of the Earth, creating a profound winter in a previously tropical world. The tropical trees and shrubs begin to die; the creatures that live in them or feed on them begin to die; the carnivores that depend on these smaller herbivores as food begin to die. The "middle life" of the Mesozoic Era—a time beginning 250 million years ago—comes to the end of its nearly 200-million-year reign.

Following months of darkness, the Earth's skies finally begin to clear, but the mass extinction—the deaths of myriad species—is not yet over. The impact winter comes to an end, and global temperatures begin to rise—and rise. The impact has released enormous volumes of water vapor and carbon dioxide into the atmosphere, creating an intense episode of greenhouse warming. Climate patterns change quickly, unpredictably, and radically around the globe before the Earth's temperatures regain their normal equilibrium. They swing from tropical to frigid,

*Debris from the K-T impact would have created continent-size fireworks before raining ash and darkness on the planet for years.*

then back to even more tropical than before the impact, all in a matter of a few years. These temperature swings produce more death, more extinction. The dinosaurs die out, as do most—but not all—mammals. Most life in the sea is exterminated.

The end-Cretaceous catastrophe was global, immense. It shares many characteristics with the Permian extinction so vividly exposed and expressed in the Karoo: both affected the Earth so much that they changed the nature of sedimentary rocks of the time. In France that change is clear—the latest Cretaceous rocks are green in color; the K-T boundary layer is dark mudstone, and the recovery rocks of the succeeding Tertiary are the thick pink limestone. Such changes occur only in the face of great chemical changes.

The geologist ponders the site. The boundary beds may have been a product of this single calamitous event, the impact of a huge asteroid with the Earth 65 million years ago. But the other victims on this Hendaye beach, the giant clams found in the strata beneath this site, were killed off 2 million years prior to the impact. What killed them? Was their passing (and that of many other creatures at the same time) the result of an Earth already stressed? It appears that the Cretaceous-Tertiary mass extinction, like the great Permian extinction that preceded it, was multi-causal.

The geologist's reverie is broken by a great rumbling sound, and he notices, for the first time, the giant culvert snaking down from the cliffs above, a pipe three feet in diameter, ending in the small bay he is standing in. A great deluge of brown water belches from the pipe, filling the bay with treated sewage from the plant on the bluff above. The Cretaceous rocks and the overlying Tertiary strata are quickly covered, clues to a long-ago extinction fouled with the last meals of the good people of Hendaye.

## Lessons from the Past

Mass extinctions are biological events. But they have been transformed into geologic evidence, and therein lies the problem. Turning flesh into stone means the loss of most biological information, and at best we have only the slenderest of clues to the events of that time. Even so, the transition of creatures during the two mass extinctions profiled above can teach us a great deal about how mass extinctions can affect the nature of evolution on the planet. Not only did the *composition* of the fauna (and flora) change radically, through the replacement of one suite of species with another, but so too did the body types of the animals and plants involved. There was not only a turnover, but also what we might call a "changeover."

The last assemblage of vertebrate animals present on Earth immediately prior to the great Permian extinction was entirely made up of four-legged types—quadrupeds. All of the dicynodonts walked on four legs, as do the majority of reptiles and mammals living on Earth today. By the very end of the Permian many held their legs beneath the body, as all mammals do today. Some of these types of animals survived. In the Triassic Period, soon after the mass extinction, the survivors and the earliest of the new species to evolve also could be typified as quadrupeds. But from that point on things began to change. With the first appearance of dinosaurs in the Triassic Period, a new suite of forms made its appearance: bipeds. While there were indeed many four-legged dinosaurs, the dominant form of the Mesozoic, exemplified by the allosaurs, tyrannosaurs, iguanodons, and duck-billed dinosaurs, was bipedal. Even the four-legged dinosaurs (such as the giant sauropods, stegosaurs, ankylosaurs, and ceratopsians) had body forms different from anything found among the late Permian faunas, for nothing in the late Permian had the long tails or giant sizes found among the dinosaurs. The large animal life on either side of the great Permian extinction is quite dissimilar. The body forms of the Paleozoic land life do not closely resemble those of the dinosaurian fauna that followed.

Would dinosaur body types have evolved even if the Permian extinction had not occurred? This is an unanswerable question, but we do know that the mammal-like reptile faunas of the late Paleozoic were moving toward the mammalian condition. Some investigators even interpret them as having been quite mammalian. In the absence of dinosaurs, would these same animals have produced *T. rex* or *Triceratops* clones, with body shapes mimicking those of the dinosaurs? It seems doubtful, for true mammals have never really explored the bipedal or long-tailed body types, kangaroos and some small rodents being among the few exceptions to this rule. We are left with a powerful observation: entirely new types of body forms may be the legacy of a mass extinction.

The world in the aftermath of the Permian extinction was desolate, and not only on land. In the seas the extinction was equally devastating. As on land, the great dying in the Permian seas radically reset the evolutionary agenda. Perhaps the most telling evidence of the extinction's severity is found in the western United States, in the reddish strata deposited in shallow seas following the extinction. Such warm, sunny seaways are today the sites of rich communities of organisms living above, on, and in their sandy bottoms. Because the continent of North America was farther south 250 million years ago, the shallow seaways of its western portions were in equatorial latitudes, and prior to the great mass extinction they were homes to rich and diverse coral reefs—among the most diverse habitats on Earth, then as now. Yet after the extinction, these same geographic sites were virtual biological deserts, barren of all life save a scattering of rare invertebrates and vertebrates. The most common organisms were stromatolites, layered algae that had almost disappeared from Earth more than 500 million years ago for a simple reason: with the rise of herbivorous animals, such layered mats of vegetation could not survive the incessant grazing that resulted. After the extinction, however, stromatolites made a comeback, suggesting that most of the seas were without their usual assortment of herbivores. The seas, like the land, remained impoverished for several million years. The old order passed away; the world of mammal-like reptiles and trilobites, spiky archaic trees, and gorgonopsian predators crumbled, to be replaced by a world of dinosaurs and pines, and ultimately by flowering plants and burrowing clams and bony fishes in the sea.

Eventually, the Mesozoic biota rose up for the Permian ashes, and then it too was struck down in a second great mass extinction. Across the globe, in every ecosystem, the changeover in fauna was spectacular—just as it was in the earlier Permian extinction. Ammonites and their legions of shelled cephalopod relatives disappeared from the seas, to be replaced by bony fishes and a new type of cephalopod—the cuttlefish. The reefs of the time died out, and when reefs eventually reappeared, they were composed of framework-building organisms of entirely different types. The changeover on land is far better known: the complete extinction of the dinosaurs allowed the rise of the many types of mammals we see today. And like the earlier Permian event, the enormous catastrophe ending the Mesozoic was followed by the rise to dominance of evolutionary dynasties quite different from those that came before. The lesson of these two great mass extinctions seems clear: extinction leads to evolutionary innovation. But is this always the case, and is it the only, or even the most important, lesson to be learned from such past global catastrophes?

As it turns out, these two mass extinctions were discovered by accident. In the eighteenth and nineteenth centuries it became imperative to devise some way of determining the age of rocks on the Earth's surface. By the early 1800s European and American geologists had begun to use fossils as a means of subdividing the Earth's sedimentary strata into large-scale units of time. In so doing they made an unexpected discovery: they found intervals of rock characterized by sweeping changes in fossil content. Setting out to discover a means of calibrating the age of

*Although their turn would come, mammals did not take a prominent place in the Mesozoic food chain.*

rocks, they discovered a means of calibrating the diversity of life on Earth. And they found intervals of biotic catastrophe, which were named mass extinctions.

The two largest mass extinctions—those examined above—were so profound that they were used by John Phillips, an English naturalist, to subdivide the stratigraphic record—and the history of life it contains—into three large blocks of time. The Paleozoic Era, or time of "old life," extended from the first appearance of skeletonized life 530 million years ago until it was ended by the gigantic Permian extinction of 250 million years ago. The Mesozoic Era, or time of "middle life," began immediately after the Permian extinction and ended with the Cretaceous-Tertiary extinction 65 million years ago. The Cenozoic Era, or time of "new life," extends from that last great mass extinction to the present day. At the time of Phillips's work, in the middle part of the nineteenth century, the notion that a species could go extinct was still quite new, and his recognition that not only single species, but a majority of species, could and did go extinct in short intervals of time was radical for its day.

John Phillips's 1860 paper also marked the first serious attempt at estimating the diversity, or number of species, present on the Earth in the past. Phillips showed that, over time, the diversity of life on Earth has been increasing, in spite of the mass extinctions, which are short-term setbacks in diversity. The mass extinctions somehow seemed to make room for larger numbers of species than were present before. Far more creatures were present in the Mesozoic than in the Paleozoic, and then far more again in the Cenozoic. But the mass extinctions did more than just change the number of species on Earth. They also changed the *makeup* of the Earth.

Mass extinctions are thus one of the most significant of all evolutionary phenomena. The wholesale destruction of animal and plant species in such large numbers opens the floodgates of evolutionary innovation. Far more than initiating the simple formation of new species a few at a time, the violent cataclysms of mass extinction reset the evolutionary clock. The two events profiled in this chapter are only the most severe of more than fifteen such episodes during the last 500 million years and, not coincidentally, the most consequential in bringing about new evolutionary innovation. They literally changed the course of life's history on this planet. Had the Permian extinction not taken place, there probably would have been no Age of Dinosaurs, and mammals might have dominated the planet by 250 million years ago, rather than 50 million years ago. Did this extinction delay the rise of intelligence by 200 million years? And, in turn, if the dinosaurs had not been suddenly killed off following an asteroid collision with the Earth 65 million years ago, there probably would have been no Age of Mammals, since the wholesale evolution of mammalian diversity took place only after the dinosaurs were swept from the scene. While dinosaurs existed, mammals were held in evolutionary check. Mass extinctions are thus both instigators of and obstacles to evolution and innovation. Yet much of the research into mass extinctions suggests that their disruptive properties are far more important than their beneficial ones.

On every planet, sooner or later, a global catastrophe can be expected that could seriously threaten the existence of animal life or wipe it out altogether. Earth is constantly threatened by planetary catastrophe, mainly by the comets and asteroids that cross its orbit, but potentially from other hazards of space. Yet it is not only the hazards of outer space that threaten the diversity of life on this planet—and surely on other planets. There are Earth-borne causes of catastrophe as well as extraplanetary causes.

## Do Causes Matter?

In one way or another, all mass extinctions appear to be immediately caused by changes in the "global atmospheric inventory"—changes in the components of the Earth's atmosphere or in their relative amounts. Such changes can be caused by many things: asteroid or comet impact, releases of carbon dioxide or other gases into the oceans and atmosphere during flood basalt extrusion (when great volumes of lava flow out onto the Earth's surface), degassing caused by the exposure of oceanic sediments rich in organic material during sea level changes, or changes in ocean circulation patterns. The killing agents arise through changes in the makeup and behavior of the atmosphere or in factors, such as temperature and circulation patterns, that are dictated by properties of the atmosphere. Sudden climate change was probably involved in the Permian extinction, and the Yucatán asteroid impact is the probable cause of the Cretaceous catastrophe. But there may be a further cause of major mass extinction: the emergence of a global intelligence.

The causes listed above all derive from a single source. Yet the history of mass extinction on this planet suggests that more that a single cause is associated with the events we find in the rock record. Sometimes these multiple events occur at the same time; sometimes they are separated by hundreds of thousands of years. Perhaps one perturbation stresses the planet, making it more susceptible to the next. Both the Permian and end-Cretaceous calamities appear to have been brought about by more than a single cause.

But is "cause" really such an important thing to know? Generations of humans have been inculcated with the notion that each crime must be solved, the "who" in a "whodunit" revealed. For the mass extinctions, we may have to be satisfied with understanding the effect, rather than the cause.

## The Anatomy of a Mass Extinction

The typical sequence of events in a mass extinction begins with the extinction phase, when biotic diversity falls rapidly. During this time, the extinction rate (the number or percentage of taxa going extinct in any time interval) far exceeds the "origination rate" (the number of new taxa evolving through speciation). After some period of time, the extinction phase ends and is succeeded by a second phase, often called the survival phase. This is a time of minimal diversity, but no or few further extinctions. During this interval the number of species on Earth levels out, neither increasing or decreasing. The third phase, called the rebound phase, is when taxonomic diversity slowly begins to increase. The final phase is the expansion phase, and it is characterized by a rapid increase in diversity due to the evolution of new species. The latter three phases are grouped together into what is known as a "recovery interval," which is followed by a long period of environmental stability (until the next mass extinction). The rate of the recovery is usually proportional to the intensity of the extinction that triggered it: the more intense the mass extinction, the more rapid the rate of new species formation.

Three types of taxa are generally found immediately after the mass extinction: survivors, or *holdover taxa; progenitor taxa,* the evolutionary seeds of the ensuing recovery; and *disaster taxa,* species that proliferate immediately after the end of the mass extinction. All three types of taxa are generally forms that can not only tolerate, but thrive in, the harsh ecological conditions following the mass extinction event. They are generally small, simple forms capable of living and surviving in a wide variety of environments. We have another term for such organisms: weeds.

The recovery interval is marked by a rise in diversity. This sudden surge in evolution is generally due to the many vacant niches found within the various ecosystems following the mass extinction. Because so many species are lost in a mass extinction, it creates new opportunities for speciation. Darwin once likened the speciation process to a wedge: the modern world has so many species in it that for a new species to survive and compete, it must act like a wedge, pushing out some other already entrenched species. But after a mass extinction no wedging is necessary.

Early on, virtually *any* new design will do. Many new species appear with morphologies or designs seemingly rather poorly adapted to their environment and inferior to those of species existing prior to the extinction. Rather quickly, however, a winnowing process takes place through natural selection, and new, increasing efficient suites of species rapidly evolve.

The great mass extinction ending the Permian created a long-term deficit in diversity, but eventually, in the Mesozoic Era, that deficit was made up. In fact, after every mass extinction that has occurred on Earth over the past 500 million years, biodiversity has not only returned to its former value, but exceeded it. Sometime during the last 100,000 years, biodiversity appears to have been higher than it has been at any time in the past 500 million years. If there had been twice the number of mass extinctions, would there be an even higher level of diversity than there is on Earth now?

Interesting as this question is, it has not yet been tested in any way. The fossil record, however, does yield some evidence that mass extinctions belong on the deleterious rather than the positive side of the biodiversity ledger. Perhaps the best such clue comes from the comparative history of reef ecosystems. Reefs are the most diverse of all marine habitats; they are the rainforests of the ocean. Because they contain so many organisms with hard skeletons (in contrast to a rainforest, which bears very few creatures with any fossilization potential), we have an excellent record of reefs through time. Reef environments have been severely and adversely affected by all past mass extinctions. They suffered a higher proportion of extinctions than any other marine ecosystem during each of the six major extinction episodes of the last 500 million years. After each mass extinction reefs disappear from the planet, and usually take tens of million of years to become reestablished. When they do come back, they do so only very gradually. The implication is that mass extinctions, at least for reefs, are highly deleterious and create net deficits of biodiversity. And whether we are talking about reefs, rainforests, or any other ecosystem, the reality is that for millions of years following a mass extinction the biodiversity of the planet is impoverished.

So, while there are many who would argue that since mass extinctions are sources of innovation, a modern one would not be such a bad thing, as it would be the source, ultimately, of a new age and even greater biodiversity, I will argue in the following pages that this is simply not the case.

*This arsinotherium, a distant relative of the rhinoceros, contemplates its geological past.*

TWO

# THE NEAR PAST

## The Beginning of the End of the Age of Megamammals

We are more dangerous than we seem and more potent in our ability to materialize the unexpected that is drawn from our minds.
—LOREN EISELEY, *The Unexpected Universe*

Far inland from Cape Town in South Africa, the high rocky ramparts of what is known as the Great Escarpment have dried the air and created a desert. This region is now home to many sheep and a few towns. The largest of the latter is Graaf Reinett, the self-styled jewel of the Karoo. Graaf Reinett is surrounded by high "koppies" of sedimentary rock, and its outskirts are ringed by shanties and so-called game reserves, large vacant tracts of thorn and scrub. The town itself is indeed like an emerald on brown dirt; it is a green oasis surrounded by the dusty parchment of the Great Karoo Desert, a town kept verdant by an encircling river providing life-giving water. Graaf Reinett itself is now a haven for tourists, for it is a virtual museum of nineteenth-century Afrikaans architecture, a melding of Dutch, German, and Huguenot influences amid blooming gardens and staid tree-lined streets. Tree-lined, that is, in the "White" part of town. There are few trees and little green in the nearby township to which the region's blacks are relegated.

The largest hotel in town is the Drosty, a picturesque assortment of stone cottages lining a cobbled lane and two restaurants serving the best meals in the Karoo. The Drosty has been restored to the look of its glory years, the late nineteenth century; each room is filled with antiques, and the staff is dressed to match. The ancient bar is wood and memory. Old photos line the walls, scenes of the town taken in the mid- to late nineteenth century. One of the photos shows elephant tusks piled high in the street in front of the newly built hotel.

On my first visit to the bar I came upon this photo and asked the ebony and venerable barman where the tusks had come from. There must have been hundreds of them in the huge pile, and it is clear from the photo that a brisk trade of some sort was going on around them. An auction, perhaps. The old bartender looked at the photo, as if for the first time, and professed ignorance; all he knew was that there had never been elephants living around Graaf Reinett in his tribe's memory.

The Karoo is dry and dusty; it does not seem like elephant country, so I believed him. But as years went by and I learned more about elephants, I began to wonder. Elephants can and do live in places far drier than the Karoo—the Kalahari Desert, for instance. By Kalahari standards, the Karoo is a verdant paradise. Elephants are consummate imperialists, and were once found on five continents. Why not the Karoo?

Each time I returned to the Karoo I asked local people—white and black—if they had ever heard of elephants in the region. I always heard the same story: there had never been elephants in the Karoo. But *never* is a taboo word for a paleontologist used to dealing in millions of years, and I inquired further, coming, at last, to the door of James Kitching, a retired professor.

Kitching is a fellow paleontologist, born in the Karoo, who then went on to great fame as one of the world's most celebrated bone hunters. In the 1960s he made what might be the most important fossil find of the twentieth century. On a cold, rock-strewn slope in Antarctica he discovered a specimen of the mammal-like reptile *Lystrosaurus.* This same creature is perhaps the most common vertebrate fossil in the Karoo—so common, in fact, that Kitching no longer bothers to collect them. But this particular fossil, the first common animal of the Triassic period, had never been recovered in Antarctica, and its discovery there constituted a powerful geologic proof that, 250 million years ago, Antarctica and Africa lay joined. In fact, at that time, all of today's southern continents were united in a single "supercontinent" named Gondwanaland, whose components—Africa, India, South America, and Antarctica—subsequently split apart and drifted across the Earth's surface like great stately cruise ships, carrying their animals—and fossils—with them. Kitching's find of *Lystrosaurus* in Antarctica constituted one of the proofs of what is now regarded as fact: that continents drift.

My real purpose for coming to see Kitching was to discuss Permian fossils, but I soon asked him about the elephants as well. He laughed dryly. "Of course there were elephants here. They followed the watercourses up from the coast, sticking to the rivers and finally arriving here in the Karoo. I have come across their bones many times around here. The last were killed off about the turn of the century."

I still remember that phrase, "killed off." Not surprisingly, the great pile of tusks in the photo at the Drosty Hotel in Graaf Reinett came from local elephants, hunted to extinction by the local farmers and townspeople. But what struck me was not that the local elephants had been killed off, for extinction is a fact of life, but that even the *memory* of their existence had been killed off in less than a century since the last one died. They were hunted to extinction and then forgotten. Where once the great elephants roamed in great herds, nothing is left of them but a fading photograph. No longer even a memory, they are now a part of a vanishing Africa, where wilderness has been transformed into farmland in a single generation.

Africa is revered for its abundance of large mammals. Nowhere else on Earth can such diversity of large herbivores and carnivores be found. Yet this animal paradise—instead of being the exception—was once the rule: all of the world's temperate and tropical grazing regions were quite recently of African flavor. But the elephants of the Karoo are just one casualty of an extraordinary event that has depleted the Earth's biodiversity of large mammals over the past 50,000 years. Is this a mass extinction? Are the forces causing it still under way? Or was it the first cause in a multi-causal event now entering a new phase?

Although the disappearance of large animals poses a tremendous challenge to those studying extinction, one significant lesson we can take from the past is that the extinction of large animals has a far more important effect on the structure of ecosystems than does the extinction of smaller ones. The extinction at the end of the Cretaceous was significant not because so many small mammals died out, but because the dinosaurs did. It was the removal of these very large land-dwelling animals that reconfigured terrestrial environments. In similar fashion, the removal of the majority of large mammal species across most of the world over the last 50,000 years is an event whose significance is only now becoming apparent, and one that should have lasting effects for additional millions of years into the future.

In the late Pleistocene Epoch, at the end of the Ice Age, about 15,000 to 12,000 years ago, a significant proportion of the large mammals in North America went extinct. At least thirty-five genera (and thus at least that many species) disappeared from North America during this time. Six of these lived on elsewhere (such as the horse, which died out in North and South America but lived on in the Old World); the vast majority, however, died out utterly. The lost species represented a wide

*The arrival of humans in North America was one of the most devastating events ever to occur on the continent.*

spectrum of taxonomic groups, distributed across twenty-one families and seven orders. The only unifying characteristic of this rather diverse lot is that most (but certainly not all) were large animals.

The best-known and most iconic of these lost species were the elephant-like animals—the probisiceans. They included mastodons and gomphotheres as well as mammoths, which were closely related to the two types of still-living Old World elephants. Of these, the most widely distributed in North America was the American mastodon, which was found from coast to coast across the unglaciated parts of the continent. It was most abundant in the forests and woodlands of the eastern part of the continent, where it browsed on trees and shrubs, especially spruce trees. The gomphotheres, a bizarre group quite unlike anything now alive, are questionably recorded from deposits in Florida, but otherwise were widely distributed in South rather than North America. The last group, the elephants, was represented in North America by the mammoths, comprised of two species, the Colombian mammoth and the woolly mammoth.

The other group of large herbivores iconic of the Ice Age in North America was the giant ground sloths and their close relatives, the armadillos. Seven genera constituting this group went extinct in North America, leaving behind only the common armadillo of the American Southwest. The largest animals of this group were the ground-living sloths, ranging from the size of a black bear to the size of a mammoth. An intermediate-sized form is commonly found in the tar pits of present-day Los Angeles, while the last and best known, the Shasta ground sloth, was the size of a large bear or small elephant. Also lost at this time were the North America glyptodont, a heavily armored creature 10 feet in length, and an armadillo, a member of the genus represented today only by the common nine-banded armadillo.

Both even-toed and odd-toed ungulate animals died out as well. Among the odd-toed forms, the horse, comprising as many as ten separate species, went extinct, as did two species of tapirs. Losses were even greater among the even-toed ungulates. Thirteen genera belonging to five families went extinct in North America alone in the Pleistocene extinction, including two genera of peccaries (wild pigs), a camel and two llamas, the mountain deer, the elk-moose, three types of pronghorns, the saiga, the shrub ox, and Harlan's musk ox.

With so many herbivores going extinct, it is no surprise that many carnivores also died out. These included the American cheetah, a large cat known as the scimitar cat, the saber-toothed tiger, the giant short-faced bear, the Florida cave bear, two types of skunks, and a canid.

*A few of the local fauna from Los Angeles, circa 18,000 B.C., courtesy of the La Brea tar pits.*

Finally, some smaller animals round out the list, including three genera of rodents and the giant beaver. But these were exceptions—most of the animals that died out were large in size.

The animal extinction in North America coincided with a drastic change in plant community makeup. Vast regions of the Northern Hemisphere went from being made up primarily of highly nutritional willow, aspen, and birch trees to far less nutritious spruce and alder groves. Even in those areas dominated by spruce prior to the extinction, a diverse assemblage of more nutritious plants was still available. But as the number of nutritious plants began to decrease due to climate change, herbivorous mammals would have increasingly foraged on the remaining more nutritious plant types, thus exacerbating their demise. The reduction of their food supplies may, in turn, have led to reductions in size for many mammal species. As the Pleistocene ended, the more open, higher-diversity spruce forests and nourishing grass assemblages were rapidly replaced by denser forests of lower diversity and lower nutritional value. In the eastern parts of North America the spruce stands changed to large, slow-growing hardwoods such as oak, hickory, and southern pine, while in the Pacific Northwest great forests of Douglas fir began to cover the landscape. These forest types have a far lower carrying capacity for large mammals than the Pleistocene vegetation that preceded them.

It was not just North America that suffered such severe losses. Until recently, North and South America were isolated from each other, and hence their faunas underwent quite separate evolutionary histories. Many large and peculiar mammals evolved in South America, including the enormous, armadillo-like glyptodonts as well as the giant sloths (both of which later migrated and became common in North America), giant pigs, llamas, huge rodents, and some strange marsupials. When the Isthmus of Panama formed some 2.5 million years ago, free interchange between the two continents began.

As in North America, a mass extinction of large mammals occurred in South America soon after the end of the Ice Age. Forty-six genera went extinct in South America between 15,000 and 10,000 years ago. In terms of the percentage of fauna affected, the mass extinction in South America was even more devastating than that in North America.

In Australia the losses were even greater. Since the Age of Dinosaurs the Australian continent had been an isolated landmass. Thus its mammals were cut off from the mainstream of the Cenozoic Era and followed their own evolutionary path, resulting in a great variety of marsupials, many of them large. During the last

50,000 years however, forty-five species of marsupials belonging to thirteen genera were killed off. Only four of the original forty-nine large species (greater than 20 pounds in weight) present on the continent 100,000 years ago survived. Of course, no new arrivals from other continents bolstered the disappearing Australian fauna. Large reptiles also disappeared, including a giant monitor lizard, a giant land tortoise, and a giant snake, as well as several species of large flightless birds. The larger creatures that did survive were those capable of speed, or with nocturnal habits.

The wave of extinctions affecting the faunas of Australia, North America, and South America coincides both with the first appearance of humanity in all three regions and with substantial climate change. Reliable evidence now shows that humans reached Australia between 35,000 and 50,000 years ago. Most of the large Australian mammals were extinct by about 30,000 to 20,000 years ago.

A different pattern emerges in the areas where humankind has had a long history, such as Africa, Asia, and Europe. In Africa, modest mammalian extinctions occurred 2.5 million years ago, but later losses, compared with those of other regions, were far less severe. The mammals of northern Africa, in particular, were devastated by the climate changes that gave rise to the Sahara Desert. In eastern Africa, few extinctions occurred, but in southern Africa, significant climate changes occurring about 12,000 to 9,000 years ago were coincident with the extinction of six species of large mammals. In Europe and Asia there were also fewer extinctions than in the Americas or Australia; the major victims were the giant mammoths, mastodons, and woolly rhinos.

**The Pleistocene extinction can thus be summarized as follows:**

- Large terrestrial animals were the primary victims; smaller animals and virtually all marine animals were spared.
- Large mammals survived best in Africa. The loss of large mammalian genera during the last 100,000 years in North America was 73%; in South America, 79%; in Australia, 86%; but in Africa, only 14% died out.
- The extinctions were sudden in each major group, but occurred at different times on different continents. Powerful carbon dating techniques allow very high time resolution. These techniques have shown that some species of large mammals may have gone completely extinct in periods of 300 years or less—a nanosecond in evolutionary time.
- The extinctions were not the results of invasions by new groups of animals (other than *Homo sapiens*). It has long been thought that many extinctions

> take place when new, more highly evolved or adapted creatures suddenly arrive in new environments. Such was not the case in the Ice Age extinctions, for in no case can the arrival of some new fauna be linked to extinctions among the forms already living in the given region.

This various evidence has suggested to many that humanity provoked the Pleistocene mass extinction. Others argue just as vigorously that the cause of the megamammal extinction was the changes in vegetation that occurred during the intense climate changes accompanying the end of the Pleistocene glaciation. In fact, most discussion about this extinction deals exclusively with this argument over humans versus climate as its cause.

For the sake of our arguments, however, the cause is irrelevant. No one doubts that whatever its cause, the Ice Age mass extinction resulted in a major reorganization of terrestrial ecosystems on every continent save Africa. But today Africa is making up for lost time, losing its megamammals as the large herds of game become restricted to game parks and reserves, where they become easy prey to poaching within their newly restricted habitats.

The end of the Ice Age megafauna is not a clearly defined line, like those drawn in the sand at the Permo-Triassic and Cretaceous-Tertiary boundaries. But then we are looking at it from the present, and, geologically speaking, it is just a moment away. At this distance, intervals of time lasting 10,000 years or less are insignificant and probably beyond the resolution of our technology—when viewed from tens to hundreds of millions of years away. The end of the Age of Megamammals looks protracted from our current vantage point, but it will look increasingly sudden as it disappears into the past—one of the odd aspects of time. But there may be more to the story. The megamammals still left on Earth now make up the bulk of endangered species, and many large mammalian species are now at risk. If the first phase of the modern mass extinction was the loss of megamammals, its current phase seems concentrated on plants, birds, and insects as the planet's ancient forests are turned into fields, cities, and toxic waste dumps.

As we race forward into the new millennium, powered by an Internet-fueled economy, biologists strain to look forward in time, watching for the suspected new biological onslaught to begin. In my view, it has already happened. It is visible in the rearview mirror, a roadkill already turned into geologic litter—bones not yet even petrified—the end of the Age of Megamammals.

*The Norway rat, one of the few mammals as successful as humans, steps off the boat in Polynesia, circa 1767.*

THREE

# INTO THE PRESENT

I have seen no grander sight than the fire upon a country which has never before been burnt.

—SAMUEL BUTLER

Oxford is the odd twin of Cambridge, the slightly less illustrious sibling, a bit of a ne'er-do-well compared with its slightly more senior fraternal twin. Immensely illustrious certainly, old, rich, and smart. But not Cambridge.

A geologist will note other differences immediately. While Cambridge sits upon the chalk of the Cretaceous, Oxford lies in the opposite direction from London, toward older rocks. Its buildings are made of yellow and tan Jurassic sandstone, limestone, and oolite (a delectable geologic term used to describe a particular grainy limestone). And for reasons unknown, it has bred or attracted a batch of evolutionists quite different from the Cambridge mix. Richard Dawkins and Robert May are there. But the most iconoclastic may be Norman Myers, a conservationist turned futurist who sees and fears the worst not only for the future of biodiversity, but for the future of evolution itself, in the upcoming years. Myers has been the most vocal prophet crying that the end of biodiversity—at least as we know it—is nigh.

Has the Earth indeed entered a new mass extinction event, or is such an event nearly over? The first of these two contentions was radical even as late as the 1980s, but in the earliest part of the twenty-first century it seems accepted as fact. (The second, that the most consequential phase of this extinction, at least for large animals, is over, is still new scientific territory.) Numerous articles and a succession of books have all treated this subject in detail. Yet Myers was there first—even arguing that the loss of the Pleistocene megamammals is connected to the modern-day biodiversity crisis. According to this hypothesis, the extinction of so many large animals (mainly mammals and birds) over the past 50,000 years was only the beginning of a larger wave of extinction continuing into the present and for some unknown period into the future as well.

Myers believes that a new phase of this mass extinction—a broadscale reduction in biodiversity—has been under way since about 1950, when a major increase in human encroachment into wildlife environments began. At that time approximately 1.7 billion people lived in the so-called developing countries, located in large tropical and semitropical regions characterized by vast forests and other undisturbed habitats for wildlife. The human population of these regions was approaching 5 billion people by the turn of the millennium. Myers maintains that the forces producing deforestation, desertification, soil erosion and loss, inefficient agriculture, poor land use, inadequate technology, and above all, grinding poverty are driving habitat destruction and, ultimately, extinctions of species, and that these forces are most pronounced in developing equatorial countries. He estimates that 50% of the world's species will go extinct over the next several centuries at most. While this estimate may sound drastic, it is in line with those of other biodiversity experts, including E. O. Wilson's 1992 estimate that 20% of all species will go extinct before 2020 and another 30% or more thereafter; Peter Raven's 1990 calculation that 50% of all species on Earth will be extinct by the year 2100, and Paul and Anne Ehrlich's 1992 estimate that 50% of all species will be extinct by 2050.

All of the aforementioned seers take the position that the majority of the modern mass extinction is soon to occur (but has not yet happened). But how accurate is this view? Where are the figures on current extinction rates to support this claim?

## Measuring Species Diversity

Determining rates of species loss seems straightforward: tabulate the number of species living at a given period of time, and compare that number to the number living at other time intervals. Yet there are numerous problems with this seemingly simple methodology. To arrive at extinction numbers, we need an accurate census of the living. Such a global census of biodiversity at the species level is still lacking.

No one disputes that the activities of humankind have caused extinctions in the recent and not so recent past. The phrase "dead as a dodo" is not pure whimsy. But there is currently great debate about the extent of anthropogenic extinctions, and even more about the prospects for such extinctions in the future. Ultimately, the entire issue devolves into numbers. But the numbers we need are very difficult to

*One hundred years ago the vast Amazon rainforest was a virtually pristine ecosystem. Today there is almost no portion of it that has not been touched by humans.*

obtain: How many species are there on Earth? How many have there been at various times in the past? How many species have gone extinct in the last millennium, the last century, or even the last decade or year? And, most important of all, how many will be gone in the next century, or millennium, or million years? None of these numbers is directly obtainable; all have to be reached, if at all, by abstraction, inference, deduction, or just plain guesswork. In contrast to the estimates above, some scientists wonder whether the loss of species will approach even 10% of current world diversity, and suggest that such a small loss would hardly be noticed.

Why is there any controversy at all about how many species are presently on Earth? In this day and age, when modern science can detect planets among stars light-years away and can deduce the age of the universe from the movement and activity of subatomic particles, what could be simpler than counting up the number of species on Earth and then, over a twenty-year time period, for instance, observing how many are going extinct? Such an endeavor would require a large army of biologists, many more than the small handful actually engaged in this type of research. In reality, we have only the haziest idea of how many species currently exist on Earth, how many there have been in the past, and how many are going extinct at any given time. It is our lack of the most basic and necessary information—the current number of species presently living on Earth—that is the cause of the greatest dissention.

Of the Earth's 1.6 million currently described creatures, about 750,000 are insects, 250,000 are plants, 123,000 are arthropods other than insects, 50,000 are mollusks, and 41,000 are vertebrates; the remainder is made up of various invertebrate animals, bacteria, protists, fungi, and viruses. The majority of organisms leave no fossil record.

The precise figure for the world's biodiversity is not known. There is no central registry for the names of organisms, and because of this, many species have been named several times. Taxonomist Nigel Stork believes that the level of synonymy may approach 20%. For example, the common "ten-spotted ladybird" found in Europe has forty different scientific names, even though it represents but a single species. Such mistakes may seem easily avoidable, but many species exhibit a wide range of variation, and the more extreme examples of a given species are often mistakenly described as new or separate species.

Does this mean that the number of species on Earth today is less than the currently defined 1.6 million? Probably not. Most biologists studying biodiversity suspect that there are far more, but an intense debate rages about exactly how many more. The most extreme estimates are in the range of 30 to 50 million species, meaning that taxonomists have named just over 3% of the species on Earth, and thus have barely begun their work in the 250 years or so since Linnaeus set out the task of describing every species. Other, more cautious souls posit much lower numbers, between 5 and 15 million species. Yet even with this lower number it is clear that the work of describing the Earth's biota has a long way to go.

The accuracy of species counts varies from group to group. For some groups, such as birds and large vertebrates, our census is nearly complete; there will be very few new discoveries of new species. Yet for the majority of invertebrate groups, and for the legions of one-celled organisms such as protozoans, bacteria, and other microbes, there are surely millions of forms yet undescribed.

It is clear that scientists will never succeed in describing every species (however wonderful that would be). Nevertheless, there is a pressing need to establish a reasonable estimate of world biodiversity. Is it closer to 1.6 million or 50 million? How can a more reliable estimate be established without describing every living species?

There have been several ingenious attempts to arrive at a reasonable estimate of the number of species living on Earth. As far back as the 1800s British zoologists knew that insects are the single most diverse group of animals on Earth and tried to census world insect diversity, coming up with an estimate of 20,000 species. It is now known that at least that many insect species are found in Britain alone. How, then, to make a more accurate accounting of the world's species? The favored method today is to use the ratio of known to unknown species in taxonomic groups that have been long studied and are considered essentially well known (such as birds and mammals) to estimate total world biodiversity. Botanist Peter Raven used this method in 1980 to suggest that world biodiversity is about 3 million species. Specialists on insect diversity have been particularly adept at coming up with new and clever ways of making such estimates. Nigel Stork and his colleague K. G. Gaston noted that of 22,000 insect species known in Britain, 67 are butterflies. Assuming that the ratio of butterflies to other insect species is the same in the rest of the world (an assumption utterly untested, but plausible), they arrived at a global biodiversity estimate of 4.9 to 6.6 million species of insects alone.

A second method of arriving at a global biodiversity estimate is to extrapolate from samples. Samples from a particular geographic area, rather than a taxonomic group, are scaled upward to encompass the entire biosphere. It was this method that yielded the most famous of all recent biodiversity estimates, that of Smithsonian entomologist Terry Erwin published in 1982, which posited that there are at least 30 million species of beetles in the world's tropical forests. This particular estimate has been so widely quoted (and misquoted) that it bears further scrutiny.

Erwin's concern at the time was not global diversity, but simply that the upper reaches of the rainforest canopy were little known and rarely sampled by taxonomists. Most known organisms from rainforests came from the forest floor. The multistoried canopies of large trees were known to harbor a quite different fauna, yet because of the difficulties of sampling this environment, its inhabitants were poorly known. Erwin devised a new method of sampling the upper canopy regions. Over three field seasons he poisoned and collected all of the insects living in nineteen trees belonging to a single tree species (*Luehea seemannii,* a forest evergreen). He found that there were an average of 163 species of beetles specific to that particular tree species. So far, well and good. At this point, however, Erwin set out a number of assumptions to arrive at his famous 30 million beetle estimate. First, he assumed that each hectare of forest in his sample area contained, on average, 70 different tree species. He then assumed that each of these tree species also had its own 163 species of beetles—a veritable army of beetles specific to that tree living in its canopy. By multiplying these numbers, he arrived at an estimate of 11,140 tree-specific beetle species in one hectare of forest in Panama, and then added in another 1,038 species of beetles just passing through the trees, to arrive at a figure of 12,448 beetle species per hectare of forest. Next, he assumed that beetles made up 40% of the total arthropod fauna in the canopy, so that the entire biodiversity of insects in his one hectare was 31,120 species of arthropods. He then added another third of this total to his estimate to take into account the insects found on the forest floor, and arrived at a grand total of 41,389 arthropod species per hectare of Panamanian forest. The final step—scaling up from a hectare of Panamanian forest floor to the entire world—was accomplished in the following way. Erwin noted that there are 50,000 species of trees in the tropics. Assuming his figure of 163 host-specific beetle species per tree, he arrived at his much-repeated estimate of 30 million species of beetles in the world.

Erwin's "thought experiment" was simple and elegant, but full of untested assumptions. However, since it was based (at least at the start) on real sampling in regions that up until that time were virtually unknown, it took on a life of its own and was treated quite seriously. It is still the basis for the larger biodiversity estimates cited today.

The Erwin estimates were widely publicized, and rightly so. They gave us a whole new view of global biodiversity. But because of the way in which they were produced, they were immediately controversial. New surveys that attempted to confirm or falsify these new, higher estimates of global biodiversity soon followed. One of the most thorough was carried out in Indonesia.

Project Wallace was a yearlong collaboration between scientists of the Natural History Museum in London and the Indonesia Department of Science. Some two hundred net-waving, jar-toting entomologists descended on the island of Sulawesi. Many insects were collected, including more than 6,000 species of beetles and almost 1,700 different species of flying insects, more than 60% of which were new to science. These scientists, applying methods similar to Erwin's, estimated a worldwide biodiversity of between 1.8 and 2.6 million insects. Since insects are only one part of total biodiversity (albeit the single most important one), these new estimates confirm that world biodiversity in indeed far higher than the 1.6 million species currently described. On the other hand, even with so many insect species, world biodiversity would still fall well short of the 30 million species predicted by the Erwin estimate.

Yet another type of estimate was derived by noted biologist Robert May, who in 1988 pointed out that the observed correlation between body size and species diversity could be used to arrive at a rough estimate of world biodiversity. Using such a method, May estimated that the Earth contains between 10 million and 50 million species, a figure that seems to support the Erwin estimates.

## Genetic Losses

One of the great surprises of the mid-1960s to 1970s was the discovery that species—virtually all species—are characterized by far higher amounts of genetic variability than previously supposed. The then-nascent techniques of gel electrophoresis and DNA sequencing allowed geneticists to evaluate just how different individuals of the same species were. While everyone knew that genomes—the number and type of genes—varied tremendously from species to species, no one foresaw the great genetic variability that characterizes virtually every living species.

Every organism carries a large number of genes: a bacterium typically carries about 1,000 genes, a mushroom about 10,000, and typical higher plants and animals as many as 50,000 to 400,000. It is variability among these genes that differentiates the various species on Earth, today as in the past. But there remains a great deal of variability *within* each species as well, which creates the various "races," sub-races, and populations that make up a species. This variability appears to be of utmost service to species, for it provides a hedge against sudden changes in the environment: in highly variable populations, there will probably be at least a few individuals that are "preadapted" for whatever new conditions come along, thus allowing the survival of the race. Any reduction in genetic variability is thus dangerous to a species—and appears to be a sure indicator of a species sliding toward extinction. Prior to the final extinction of a species, we see a dying off of its populations caused by a reduction in the overall genetic variation.

In the mid-1970s, a study of twenty-four proteins extracted from North American elephant seals revealed that this rare and highly endangered species shows essentially no genetic variation. This particular species had been hunted nearly to extinction, and even though there has been a rebound in the population since it was protected from further hunting, its overall genetic makeup was severely affected. The elephant seal population is said to have passed through a "population bottleneck." At its low point, each seal seeking a mate had only a very small number of other seals to choose from, resulting in severe inbreeding and a loss of genetic variation. Such unions between close relatives are often characterized by high rates of birth defects, retardation, and reduced sperm counts. Inbreeding is so deleterious that humans of every culture have produced laws against it.

An extreme example of such genetic loss is the Florida panther. This subspecies of the American cougar has been reduced to fewer than thirty individuals in the wild. Low sperm counts and damaged sperm characterize the remaining males. Genetic studies show that this subspecies has the lowest genetic variation of any of the extant cougar populations.

## Estimates of Current Extinction Rates

It can be argued that the current mass extinction is far less calamitous than either the end-Paleozoic or end-Mesozoic events because a lower *percentage* of families and genera are going extinct now than in the past. The severity of a given extinction event is commonly tabulated as the percentage of existing taxonomic units, be they families, genera, or species, that go extinct. Using this measure, it has been argued that the extinctions that have occurred since the onset of the Ice Age have been trivial compared with the great extinctions of the Paleozoic and Mesozoic eras because the percentage of taxa that have gone extinct is but a tiny fraction of the total diversity of the Earth. What is being overlooked, however, is the fact that the absolute—not relative—number of species (or other category) that have *already* gone extinct in the last million years may be substantial. For instance, biologists Stors Olson and Helen James have published data suggesting that as many as a thousand species of birds have disappeared from the Earth in the last two to five millennia.

This represents perhaps 20% of the total bird biota on the planet. These are species that have left a fossil record, and thus species we know about. How many more have gone extinct without leaving a trace?

Extinction is the ultimate fate of every species. Just as an individual is born, lives out a time on Earth, and then dies, a species comes into existence through a speciation process, exists for a given span of years (usually counted in the millions), and then eventually becomes extinct. Thus, extinctions of species happen all the time, not just during mass extinction events. University of Chicago paleontologist David Raup calls this concept *background extinction.* The fossil record can be used to tabulate the rate of such "random" extinctions taking place throughout time, and that rate turns out to be remarkably low. Raup has calculated that the background extinction rate during the last 500 million years has been about one species every four to five years. In contrast, Norman Myers has estimated that four species *per day* have been going extinct in Brazil alone over the past thirty-five years. Biologist Paul Ehrlich has suggested that by the end of the twentieth century, extinction rates were measurable in species per hour.

If attaining a reliable estimate of global species diversity has caused problems, estimates of current extinction rates have been no less controversial. While many different people disagree strongly on the number of species on Earth, and on the rate at which these species are currently declining in number, on one issue there is no disagreement: the vast majority of species currently living on Earth are found in the tropics, mainly in rainforests.

Tropical rainforests are characterized by a high canopy, often 30–40 meters above the ground with emergent trees towering to 50 meters, and two or three separate understories of vegetation. They are complex, layered communities with enormously varied and changing environments and microclimates.

Tropical rainforests today are found in three principal regions. The most extensive is the American, or Neotropical, rainforest region, centered in the Amazon Basin but extending up the Caribbean slope of Central America to southern Mexico. The Neotropical rainforest comprises about half the global areal total, and about one-sixth of the area of all broad-leaved forests in the world. The second large block occurs in the eastern tropics and is centered in the Malay Peninsula. The third is in central Africa.

Norman Myers estimates that between 76,000 and 92,000 square kilometers of tropical forest are lost each year to logging and field clearing, and that an additional 100,000 kilometers are grossly disrupted. This means that about 1% of the world's tropical forests are disappearing each year, a rate that will lead to the complete disappearance of all tropical forests in one century, if current practices continue. Biologist E. O. Wilson, in *The Diversity of Life,* estimated the rate of tropical forest loss in 1989 to be 1.8% per year. The Food and Agricultural Organization (FAO) of the United Nations officially placed the deforestation rate at 0.5% per year in the late 1980s.

Daniel Simberloff of Florida State University analyzed all available information regarding the rate of forest destruction—data mainly derived from satellite imagery and remote sensing. He found that the tropical forests of Asia are already virtually gone. There are currently about 92,000 recognized plant species (and an unknown number of plant species waiting to be described by science) in the New World tropical rainforests, and 704 species of birds. Simberloff has calculated that between 1950 and 2000, almost 14,000 plant species (15% of the total) and 86 bird species (12% of the total) became extinct in this region. If the tropical forests of the New World become restricted to current and planned reserves and national parks, Simberloff predicts that the extinction of over 60,000 plant species (66%) and 487 bird species (69%) will occur between 2050 and 2100 A.D. Simberloff concludes that "the imminent catastrophe in tropical forests is commensurate with all the great mass extinctions except for that at the end of the Permian."

## Recent Losses and Causes

Since 1600, a minimum of 113 species of birds and 83 species of mammals are known to have gone extinct. But these animals are large vertebrates, which through time have had a far lower background extinction rate than 5 per year. About three-quarters of these extinctions took place on oceanic islands. Historical records also suggest that, since 1600, extinction rates for these two groups have increased by a factor of 4, to produce the current extinction rates of around 0.5% of extant birds per century and 1% of mammals per century. Extinction rates in other groups of organisms have only begun to be tracked, but they are significantly higher than the historical average. In the United States, there were twice as many species of fish (350) classified as endangered in the 1990s than there were a decade earlier.

The major factor driving species to extinction in North America (and elsewhere in the world) appears to be changes in habitat, such as those that occur through climate change, desertification, or deforestation. Habitat perturbation often causes rapid extinction of species: the drying of a freshwater lake or the final submergence of an eroding island obviously causes the immediate extinction of many species that once dwelled there. Others will die off later in time.

The number of individuals in any population of organisms is always fluctuating. There may be long-term trends toward increase or decrease, or even toward constancy, but these longer-term trends are themselves made up of shorter-term fluctuations. The fluctuations themselves have traditionally been thought to be related to environmental factors: changes in food supply, increased or decreased predation or competition; physical environmental changes such as long-term temperature change or habitat change. To understand these changes, ecologists have developed a series of equations that describe how birth and death rates—the ultimate determinants of population size—are affected by the external environment.

Just how important unpredictable fluctuations are to populations was perhaps first appreciated by Robert May. In the 1970s, May showed that population fluctuations in many species of animals and plants are not necessarily random, but instead may be an aspect of chaos, the relatively newly described phenomenon in which apparent randomness isn't random after all. Although governed by precise mathematical rules, the behavior of a chaotic system is virtually impossible to predict. It may be that some populations of organisms show wild fluctuations that are caused not by external conditions, such as climate change, but by deeply rooted and complex dynamics *within* the ecosystems in which they reside. May also showed that the geographic distribution of organisms may be related to factors other than the external environment. May and his colleagues showed that population fluctuations within a patchy (or irregular) distribution may not be related simply to the favorability of each patch, but might be far more complex.

All of these findings have profound implications for conservation biology—and for the understanding of mass extinctions. In their 1996 book, *The Sixth Extinction,* Richard Leakey and Roger Lewin point out that

> the world of nature is not in equilibrium; it is not a "coordinated machine" striving for balance. It is a more interesting place than that. There is no denying that adaptation to local physical conditions and such external forces as climatic events helps shape the world we see. But it is also apparent that much of the pattern we recognize—both in time and space—emerges from nature herself. This is a thrilling insight, even if it means that the work of conservation management is made more difficult. It was long believed that population numbers could be controlled by managing external conditions (as far as possible). This must now be recognized as no longer the feasible option it was imagined to be.

Birds are relatively large and highly conspicuous members of the planet's biota, and thus are among the best-known groups in terms of both their current diversity and their history of recent extinctions. Because of this, they play a prominent role in our study and understanding of biodiversity loss. Birds also have the potential to leave a fossil record of themselves, so that diversity levels and losses in the past can be measured.

In 1997, David Steadman, curator of birds at the Florida Museum of Natural History, summarized losses of bird species since 1600 on both continents and islands. Steadman posited that humans have caused extinctions of birds in four main ways: direct predation (hunting, gathering eggs, or removing nestlings for captive breeding and pets), introduction of non-native species deleterious to bird survival, the spread of disease, and habitat degradation or loss. Of the approximately 10,000 species of birds on Earth today, about one-fourth have restricted breeding ranges (designated as 50,000 square kilometers or less). These are the species most susceptible to extinction.

Little is known about the prehistoric human impact on birds in most continental regions, but North America is one exception. Soon after humans arrived in North America, about 11,000–13,000 years ago, between twenty and forty species of birds went extinct. All of these birds may have been tied into ecosystems dependent on the large mammals that also went extinct at that time. It is likely, then, that the birds' extinctions were only indirectly tied to human causes. From 11,000 years ago until 500 years ago, only two additional bird species went extinct in North America. Since the arrival of Europeans approximately 500 years ago, an additional five to seven birds species have gone extinct, with five of these extinctions occurring in the last 200 years (the great auk, Labrador duck, passenger pigeon, Carolina parakeet, and ivory-billed woodpecker). Eight more species (the California condor, whooping crane, red-cockaded woodpecker, black-headed vireo, golden-cheeked warbler, and Kirtland's warbler) are so close to extinction that only expensive, concerted captive breeding efforts (such as that taking place for the condor) will save them.

The rates of avian extinction in tropical continental regions outside of North America have been little studied. The paleontology of birds is far better known for many islands. The relatively small land areas of most islands result in smaller local populations of all organisms and, as a result, greater sensitivity to extinction. Of the 108 species of birds known to have gone extinct worldwide since 1600, 97% came from islands. Even more extinctions occurred in prehistoric times. Steadman estimated that at least 2,000 bird species, or about 20% of the total diversity of birds on Earth, went extinct on islands after human, but before European, contact. In each case the extinctions postdated the first human contact with each island.

Predicting the future of birds is no easy feat. It may be that the most susceptible and delicate species have already gone, or will soon go, extinct. Perhaps the losses we have seen so recently will be the major losses. Yet many bird experts are not so sanguine, and see the ongoing cutting of the world's forests—and their replacement with agricultural fields, an entirely different type of habitat—as a factor ensuring the continued loss of bird species.

## Why the Modern Mass Extinction May Not Be as Bad as Projected

One of the great dangers facing those who attempt to prophesy is that estimates, coming from the best of intentions, may become more catastrophic than the data warrant. Extinction is an emotional issue for many of us, even (or especially) scientists, and emotion can color judgment and distort objectivity. There is a very real possibility that estimates of current extinction rates are inflated. Few studies are able to pinpoint how real the threat of elevated extinction rates really is, or how prolonged it will be. There is a tendency by some working in this field to cry doom when a much more muted response may be justified.

It is clear that the planet is in a period of elevated extinction rates. But just how elevated, compared with the period prior to the population run-up of our own species, is the most pressing question, and one that is very difficult to answer. There is a possibility that most of the consequential extinction (i.e., among the megamammals) has already occurred, and that little further reduction in the Earth's biota will accumulate over the next few centuries or millennia. Thus it may be that the estimation of losses through mass extinction is wildly overstated.

Following are several reasons why the current mass extinction may be less severe than many estimates predict:

1. Most species are resilient—more resilient than previously thought

For all of the extinctions currently thought to be under way, actual case histories of extinctions are rather few. Those species that have gone extinct, ranging from the dodo to the passenger pigeon, may be species that for any number of reasons were extremely susceptible to extinction to begin with. Extinction requires the death of every living individual of a given species. All species are the result of a long period of evolution. They do not just go away; something must eventually kill them all off, and that cause must be sufficient to end a history in most cases counted in millions of years.

**2.** Conservation efforts will be more successful than previously thought

Worldwide conservation efforts have brought to light the plight of many endangered species. Virtually every country on Earth now practices some form of conservation, be it by preserving large national parks or by protecting individual species or given habitats. These efforts have occurred only in the last two to three decades on a worldwide basis. Yet they have already registered a number of remarkable successes, notably in the restoration of whale and large bird species. Bans on dangerous chemicals such as DDT have vastly aided this process. These efforts alone may be sufficient to reverse the course of the oncoming and ongoing mass extinction.

**3.** Extinction rates have been overestimated

As we have seen, one of the most maddening aspects of biodiversity studies is our very poor knowledge of the most basic baseline figure, the actual number of species on Earth, and the corollary to that figure, the reduction of species numbers among various taxonomic groups and specific habitats. In very few other avenues of science are the error ranges quite so great: an order of magnitude separates the high and low figures. It may be that there are a very large number of species on Earth, and that a relatively low percentage of them have recently undergone extinction, or will do so in the near future.

## Why the Modern Mass Extinction May Be Worse than Projected

Several factors could adversely affect the biological diversity of the Earth and serve to amplify the current rate of extinction. If we accept that the current levels of extinction are related to the activities of humankind—most importantly, the conversion of previously undisturbed habitat such as rainforest or native grassland into agricultural areas—then anything causing an increase in such conversion should adversely affect the biodiversity baseline. Any reduction in the land currently available for agriculture would be likely to spur more conversion. This could happen in several ways:

**1.** Sudden climate change

Several types of climate change could reduce the current area of farmland and hence create pressure for further habitat conversion. Global warming due to the greenhouse effect could cause the tropical regions to increase in size. This in turn would cause an expansion of the desert areas at higher latitudes, producing an adverse effect on the grain belts located there. If grain regions migrate to higher latitudes in turn, they will suffer shorter and harsher growing seasons, and thus reduced yields.

A second and opposite effect would be a return to a new glacial interval. The current warm period is but an interglacial interval in a long pattern of glacial cycles that has been operating for more than 2 million years. If past patterns are any guide, some thousands of years from now ice sheets will again begin to grow and cover vast regions of the Earth in some of the most productive agricultural regions.

**2.** Sea level change

Disruption of agriculture could also come from a rise in sea level. Even small rises in global sea level will result in significant land reductions in agricultural regions, and such small-scale rises will come about if current global warming patterns continue. River deltas, for example, are among the richest of all agricultural regions, and the first to be inundated by any rise in sea level. New evidence gathered from a study of Antarctic glaciers in 2001 indicates that the rate of sea level rise may be three or four times faster than the worst-case scenario of the late 1980s and early 1990s. There may be a 20-foot sea level rise in the next two centuries.

**3.** Greater than expected human population increases

As we will see in greater detail in a subsequent chapter, the number of humans on Earth greatly affects the rest of its biota, and surely extinction rates as well. If the human population reaches some of the more extreme estimates over the next few centuries—over 50 billion people, for instance—there will certainly be greatly elevated extinction rates.

The Earth currently has more species than at any time during previous geologic epochs. This general pattern of increase in diversity over time may not continue, however. How it might change is described in the next chapter.

*Even a completely degraded environment can be successfully exploited by certain species—but others are sure to perish.*

FOUR

# REUNITING GONDWANALAND

> Be fertile and multiply, fill the Earth and instill fear and terror into all the animals of the Earth and birds of the sky.
>
> —GOD, *in a conversation with Noah*

In the now far-off decade of the 1960s a famous bumper sticker graced many a Volkswagen bus: *Reunite Gondwanaland.* In those days the theory of plate tectonics (also known as continental drift) was still in its infancy, and the waggish slogan was a cry to bring back all of the southern continents into the single continental landmass that existed at the end of the Paleozoic era, some 250 million years ago. In a strange way, that call has been heeded: Gondwanaland has been reunited. Not in any physical way—Africa is not measurably closer to Australia or South America than it was thirty years ago. But functionally they have been brought together as barriers to biotic exchange between them have been eliminated. The common travel of boats and ships across the oceans has shrunk those oceans by giving the animals and plants of the now separated continents access to their age-old corridors of dispersal. When Gondwanaland existed, it was a time of greater global homogeneity, of fewer ecological niches, of fewer and lower mountain ranges, of more uniform global climate—and, because of these factors, it was characterized by *far* lower planetary biodiversity than the present era. The functional reuniting of Gondwanaland may take us back to a lower global biodiversity reminiscent of that bygone age. This renewed homogenization of the world's biota may set the current mass extinction apart from all such previous events, for after this event there may not be a subsequent diversity increase. The planet may well stay at the low levels typical of a single landmass, rather than the higher diversity of numerous separated continents.

Biological diversity is so commonplace to us that it is taken for granted. Yet the factors leading to diversity are still great biological enigmas. Since the Cambrian Period more than 500 million years ago, the diversity of species on Earth has been fluctuating, but increasing overall. Will it continue to do so? Here I will argue that the new mass extinction, which is causing a dramatic decrease of diversity on Earth, will *not* be followed by a renewed burst of diversification, or even a return to pre-extinction diversity values until, perhaps, many millions of years have gone by.

## Trends in Diversity

What controls the diversity of a given region? How can a coral reef be so rich in life and a sand bed beside it so poor? And if we change scale, how can a large region be species-rich and a neighboring province species-poor? If we define diversity as the number of species present in any given area, can we arrive at some rough mathematical rule governing diversity? There are no simple answers to these questions. Many factors enter into the equations, such as nutrient availability, habitat type, and amount of water; there are also numerous factors affecting the formation of species, such as rates of barrier formation, rates of genetic change, and, especially, rates of extinction.

Biologists have long recognized that diversity appears to be roughly related to habitat size, and this makes good sense: the larger the area of habitat available, the more animals and plants, and at the same time, the more *different kinds* of animals and plants, can be accommodated. But is extinction rate also related to habitat size, in some inverse way? Do larger habitats or only larger population sizes protect individual species from extinction?

Some rough rules of thumb about this relationship were first formulated in the 1960s by two famous ecologists, Robert MacArthur and E. O. Wilson, who proposed a new theory relating diversity to habitat area. MacArthur and Wilson called their idea the equilibrium theory of island biogeography. In essence, it relates the area of habitat to the number of species present: as habitat area increases, so too do species numbers, and they do so in a predictable way. Similarly, as habitat area *decreases,* species numbers fall. Because the number of species bears such a predictable relationship to the area available, we can analyze the way in which deforestation, for example, leads to the shrinking of habitat and thus to the loss of species. This influential model was one of the seminal theories about the regulation of biodiversity through time. While it was originally designed to examine diversity on islands, models patterned on the theory of island biogeography have now been scaled upward to encompass continental and even global scales of community and evolutionary diversity.

MacArthur and Wilson's equations can be used to predict rates of extinction. They found, for example, that an island always has *fewer* species than a mainland or continental habitat area of the same size, even if the habitats are otherwise exactly identical. The implications of this finding are that parks and reserves, which essentially become islands of habitat surrounded by disturbed areas, will always suffer a loss of species. It also means that subdividing any sort of larger habitat into smaller patches of disturbed and undisturbed regions will increase the rate of extinction.

With these implications in mind, paleontologist Michael McKinney of the University of Tennessee has recently summarized the general traits of global diversity:

1. Diversity (which can be defined here as the number of species in the habitat being examined, be it an island, a given community, or the Earth as a whole) fluctuates around some mean equilibrium value when viewed over a time scale we might call *ecological time:* tens to at most hundreds of years. Sometimes it drops, sometimes it rises, but generally it can be considered stable.
2. Although this mean equilibrium value of diversity remains approximately constant, the component species can and do change. Local extinction, immigration, and the formation of new species drive these changes.
3. If the same system is viewed over geologic time (thousands to millions of years), the mean equilibrium value of diversity changes as forces such as continental drift or mass extinction alter the major habitats on Earth.
4. The equilibrium value of diversity is determined by competition for available and finite resources. As the total number of species increases, this competition increases, reducing the rate of formation of new species and increasing the extinction rate.

The causes of change in diversity have been debated for over a century. Generally, the proposed causes fall into two categories: abiotic factors (those brought about by nonliving aspects of the environment, such as climate change) and biotic factors (those brought about by life itself, such as competition, predation, and disease). Not surprisingly, ecologists have stressed the importance of biotic factors, viewing the world in short time spans and at the limited geographic scales of individual habitats and ecosystems. Those who examine global biodiversity from the perspective of a larger and longer framework (such as paleontologists) have long believed that abiotic changes are the most important factors determining diversity. According to this view, the two most important mechanisms regulating diversity are the rate of origination of new species and the rate of species extinction. These two competing factors are affected by abiotic factors and by each other.

In terms of the shorter ecological time scale, speciation is always a relatively slow process, while extinction can be either fast or slow. In the present-day Age of Humanity, it appears that the large-scale environmental changes causing the observed rise in extinction are abiotic—climate change and changes in landscape and vegetation—yet their ultimate cause is biotic—the actions of humans. These circumstances have no precedent on Earth.

Our understanding of the rate of diversification relies on the concept of niche saturation. For many decades ecologists have used the concept of a *niche* to describe how a particular species lives and interacts in its ecosystem. The niche is somewhat analogous to the profession of a species: what it eats, where it lives, what it does in its community. As more and more species either evolve in or invade a given community with finite energy resources, more and more of the available niches are filled. It may be that the overall carrying capacity of a given habitat, community, continent, or even the Earth, limits the number of available niches, and that these niches can become saturated with species, thus limiting the potential for new speciation. Human activities appear to be reducing the number of niches available, at least in terrestrial habitats. The replacement of a forest with a field, or a field with a city, reduces niche availability. Suddenly the world is a less heterogeneous place—just as it was during the time of Gondwanaland, 200–300 million years ago.

## Disturbance and Diversity

Since the *actual* number of species on Earth today is so important, knowing what controls that number is also important. Why are there not twice as many, or half as many, plants and animals? Why are there more now than during the time of Gondwanaland? Although there is an enormous scientific literature on diversity, this is a question that has perplexed biologists for nearly two centuries, and it appears that there is no easy answer. The most famous book on the topic—Charles Darwin's *On the Origin of Species*—does not even address the issue. Darwin was concerned with the transformation of *individuals,* rather than how and why new species form. Most of the more recent treatises on diversity, such as Yale ecologist Evelyn Hutchison's famous paper "Homage to Santa Rosalia, or Why Are There So Many Kinds of Animals?" do not examine the mechanisms leading to the origin of species, but simply describe the maintenance of species once they have evolved.

Nevertheless, this problem was reexamined recently in a thoughtful essay by paleontologists Warren Allmon, Paul Morris, and Michael McKinney, who attacked it in a different way. They asked how short-term environmental changes, or *perturbations,* as well as more severe and longer-lasting changes, which they called *disturbances,* affect evolution and diversity. Because humans are producing both perturbations and disturbances in copious quantities around the globe, this particular question is highly relevant to understanding and predicting possible future trends in diversity.

All organisms encounter perturbations in their daily lives. Fluctuations in temperature, food availability, rates of predatory attacks—these and a thousand other environmental changes are part of the everyday lives of all organisms. Sometimes, however, one or several of these changes are severe enough to kill off or otherwise remove a species or group of species from a given geographic area, creating a patch in space from which these organisms are now absent. Of course, what constitutes a perturbation or disturbance varies from species to species—a disturbance for a protozoan may not even be noticeable to a fish. Disturbances are thus species-specific. They can also be thought of as acting at many environmental scales, as well as many scales of time. Perhaps the most interesting for our purposes are time scales ranging from a thousand to a hundred thousand years—the intervals of time necessary for the speciation of large animals and plants.

Ecologists have long understood that there is a relationship between the degree of disturbance and the ability of nature to *maintain* diversity. Many studies of marine intertidal zones have shown that in areas of either too little or too much disturbance, few species occur. The disturbances can be both abiotic—such as a violent storm—and biotic—such as the incursion of a new predator. Both types of disturbance create patches of open space or habitat. By reducing the numbers of abundant species, they allow rare species to maintain their existence or allow new species to gain a foothold in the environment. In environments with little disturbance, diversity drops as a few species outcompete all the others and dominate the environment. In high-disturbance areas, diversity also stays low, since only a few species can maintain viable populations in the face of high mortality. Maximum diversity is found in areas that can be considered to have intermediate levels of disturbance. Such conditions allow many species to survive, but do not allow any particular species to take over through predation or competition.

On the other hand, there have been virtually no studies trying to link disturbance with speciation, or the *creation* of diversity. Allmon and his colleagues have suggested that, like the maintenance of diversity, the creation of diversity through the formation of new species may occur in regions of intermediate disturbance. Paleontologist Steven Stanley has postulated a similar model, noting that "high rates of speciation are actually promoted by less severe environmental deterioration—deterioration severe enough to elevate extinction rates to a moderate degree but not so severe as to cause wholesale extinction."

This idea has some interesting implications. It predicts that endemic species—those restricted to specific and hence small geographic regions—will encounter relatively higher levels of disturbance than more broadly adapted species, and therefore experience both higher levels of origination and extinction. These species—the specialists and types found in restricted ranges—are those that produce the largest number of new species. Yet they also have the highest extinction rates.

Global diversification remains a simple equation: origination minus extinction. The highest net rates of diversification seem to occur in animals with small body sizes, short generation times, wide distributions, and high abundances—beetles and rodents, for example. Although two of these traits—wide distribution and high abundance—seem to negate new speciation, they retard extinction even more. The net result is higher diversification than extinction.

## Compounded Disturbance and Ecological Surprise

All species have evolved in the presence of disturbance. Thus, disturbances that happen within a particular range of intensity—not too extreme—result in little long-term change in the nature, composition, and energy flow of a population, or even an ecosystem. But what of "compounded" disturbances, when major disturbances occur repeatedly at higher than normal frequencies?

This was the question posed by ecologist Robert Paine and his colleagues in a 1998 article. Paine has spent his entire research career studying intertidal organisms and has contributed fundamental discoveries about the architecture of ecosystems and diversity. According to Paine, disturbances, ranging from small-scale and

*Deforestation and fragmentation are the future—and bane—of post-industrial ecosystems.*

frequent perturbations to large and infrequent catastrophes, occur from time to time in any habitat. It is these cycles of disturbance that led to ecology's first paradigm, that of ecological succession. Disturbances often cause widespread mortality, leaving a residual assemblage of flora and fauna, which provides a legacy that subsequent processes and populations use to rebuild. Even large, infrequent disturbances do not appear to override the biotic mechanisms that structure the eventual recovery. Paine and his colleagues used the example of the catastrophic 1988 Yellowstone National Park fire, which burned nearly 40% of the park and was an order of magnitude larger than previous fires in the park region. Even a decade after this major event, there have been no ecological "surprises"; the ecosystems returning are similar to those present prior to the fire. But what if the park underwent another such fire ten years after the first, and then another a year after that? If such major catastrophes are compounded, will ecosystems return to their previous state? Paine and his colleagues argue that they will not.

Compounded disturbance can be portrayed in two ways. First, it can occur in the manner proposed in the Yellowstone fire example, in which a normal community undergoes a second (or multiple) disturbance before recovery from the first is completed. Second, a major stress can be superimposed on a community altered by some significant disturbance. Examples of the this second type of compound disturbance can be seen when fish stocks are depleted by overfishing and then subjected to some other type of large-scale disturbance. In such a case their recovery is markedly delayed, if it occurs at all. Climate change may produce the same effect: a series of major storms one after another may markedly alter ecosystems that have evolved under lower storm frequency regimes.

Paine and his colleagues note that the prime cause of compounded disturbance is human activity. The prime result is lowered diversity—a return to Gondwanaland.

## Plate Tectonics and Diversity

The studies above (and many others as well) suggest that compounded disturbance produced by humanity may have caused the equilibrium level of world biodiversity to drop. Yet there is a second and equally important factor that is taking us back to Gondwanaland: the functional removal of barriers to migration. In a way, to borrow from another hoary bumper sticker, we have indeed stopped continental drift.

One of the major influences on the equilibrium value of global biodiversity is continental configuration. When the various continents were united, there was obviously easy faunal interchange around the globe. When the continents are widely separated, however (as they are today), there is greater heterogeneity in environments, less faunal interchange, and many more species. Two hundred fifty million years ago, all the major continents were merged, and biodiversity was far lower than it is today. But by introducing non-native species across ecological boundaries and continents, humanity has found a way to functionally reunite the various continents, as least as far as gene flow is concerned.

Since the majority of the Earth's biodiversity today is found on continents (and there is no reason to believe that this relationship has changed over the last 300 million years), the processes of plate tectonics are especially important for life and its ecosystems. As continents have shifted their positions through time, they have affected global climate, including the overall albedo (the planet's reflectivity to sunlight), the occurrence of glaciation events, the pattern of oceanic circulation, and the amounts of nutrients reaching the sea. All of these factors have biological consequences that affect global biodiversity. Moreover, continental drift can help augment diversity by increasing the number and degree of separation of habitats (which promotes speciation).

Plate tectonics also promotes environmental complexity—and thus increased biotic diversity—on a global scale. A world with mountainous continents, oceans, and myriad islands is far more complex, and offers more evolutionary challenges, than a planet dominated by either land or ocean. James Valentine and Eldredge Moores first pointed out this relationship in a series of classic papers written in the 1970s. They showed that changing the positions and configuration of the continents and oceans would have far-reaching effects on organisms, causing increases in both diversification and extinction. Changes in continental position would affect ocean currents, temperatures, seasonal rainfall patterns and fluctuations, the distribution of nutrients, and patterns of biological productivity. Such changing conditions would cause organisms to migrate out of the new environments, and would promote speciation. The deep sea would be affected least by such changes, but the deep sea is the area on Earth today with the fewest species: over two-thirds of all animal species live on land, and the majority of marine species live in the shallow-water regions that would be most affected by plate tectonic movements.

The most diverse marine faunas on Earth today are found in the tropics, where communities are packed with vast numbers of highly specialized species. Not only are there *fewer* species at higher latitudes, but species composition is different from that in the tropics as well. Most species have fairly narrow temperature limits imposed by physiological adaptation, and since temperature conditions change rapidly with latitude, it's not surprising that the north-south coastlines of continents show a continuously changing mix of species.

In 1996, biologist P. Vitousek and three colleagues used mathematical modeling to project the number of mammalian species that would be expected on Earth if all of the continents were reunited into the configuration present at the end of the Paleozoic Era, some 250 million years ago. They concluded that the world would contain about half of the nearly 4,000 mammalian species present if we reunited Gondwanaland. These same authors speculated that the current transport of mammals from continent to continent is leading to an extinction rate of mammals that will yield approximately this same global biodiversity: 2,000 mammalian species.

## Aliens Among Us

Whenever a species arrives in an area where it was not previously found, there is a potential for biological change. Such invasions of new species have occurred throughout geologic time, yet the rate of invasion has vastly increased during the Age of Humanity. Today, no area on Earth is immune to such biological invasions. It is estimated that about 11% of all species now living in France have been introduced; in Australia the proportion is 10%, in Hawaii 18%, and in New Zealand more than 40%. These biological invasions are particularly marked in floral communities. There are records of about 1,200 native plant species in New Zealand, but there are now over 1,700 non-native plants living there as well. Although it could be argued that the introduction of so many non-native plant species has more than doubled the diversity of plant life in New Zealand, this is only a transitory result. Over time, many non-natives will inevitably drive the natives to extinction, causing world biodiversity to decline.

Biological invasions aided by humans have come in three major pulses. Over a period starting several thousand years ago until about 1500 A.D., human movement and migration caused the transport of plant and animals mainly in the Old World. Beginning in about 1500, however, a second phase of invasions began with the increasing contact between the Old World and the New due to European exploration and conquest, during which many Old World species were transported to the New World. The final phase began about 150 years ago with the globalization of species movement due to the vastly increased efficiency of human transport.

There have been many reasons for these species introductions. In some cases the introduced species were purposely brought to a new location to become animal or plant crops. Some were brought to serve as ornamentals or pets, while others were introduced for sport or hunting. Still others were introduced to control "pests," only to have the introduced species become even more destructive than the species they were brought in to control. Ironically, some introductions have occurred for purposes of either biotechnology or scientific research. Finally, there have been many accidental introductions from ship ballast and airplane holds, of "hitchhiking" seeds escaped from either wild or agricultural areas, and simply as side effects of habitat alteration.

The majority of introduced species do not survive. It is estimated that of a hundred introduced species, approximately ten will successfully colonize or naturalize, and only two or three will become pests. But those that do often become major problems, especially in fragile, endangered, or rare ecosystems, such as early successional habitats, ecosystems with few species, and ecosystems that traditionally have a low number of predators or grazers. The pest invaders often show a suite of similar characteristics: they have a high reproductive potential, many offspring, and generalized habits and food requirements. They can thus be characterized as "pioneer" species, in that they can colonize and flourish in a wide variety of ecosystems. They are often human "commensals"—species that thrive in the presence of humanity.

While the greatest consequences of these invasions are biological, their economic impact is not trivial. In the United States alone, it is estimated that the Russian wheat aphid causes as much as $130 million in crop damage each year, the Mediterranean fruit fly as much as $900 million, and the gypsy moth about $750 million. The boll weevil may have caused as much as $50 billion in damage to cotton crops during the twentieth century.

The ultimate effect of many invasions is extinction of native species, and examples of such extinctions abound. In 1959, in the Rift Valley of Africa, British colonials introduced a northern African fish called the Nile perch into Lake Victoria for sport fishing. The Nile perch is a voracious predator on smaller fishes. Prior to its introduction, the lake was home to over 300 species of endemic cichlid fishes. Yet by the early 1980s, when the problem was finally recognized, over half of the cichlid species in Lake Victoria had gone extinct because of the Nile perch.

Of all the factors causing the translocation of species, the exchange of ballast water may be among the most important and the most difficult to stop. Thousands of species are transported around the globe in ships' ballast water. When a ship takes on ballast water, it takes up the plankton of a given region, which often contains the juvenile stages of marine animals and plants. These organisms are

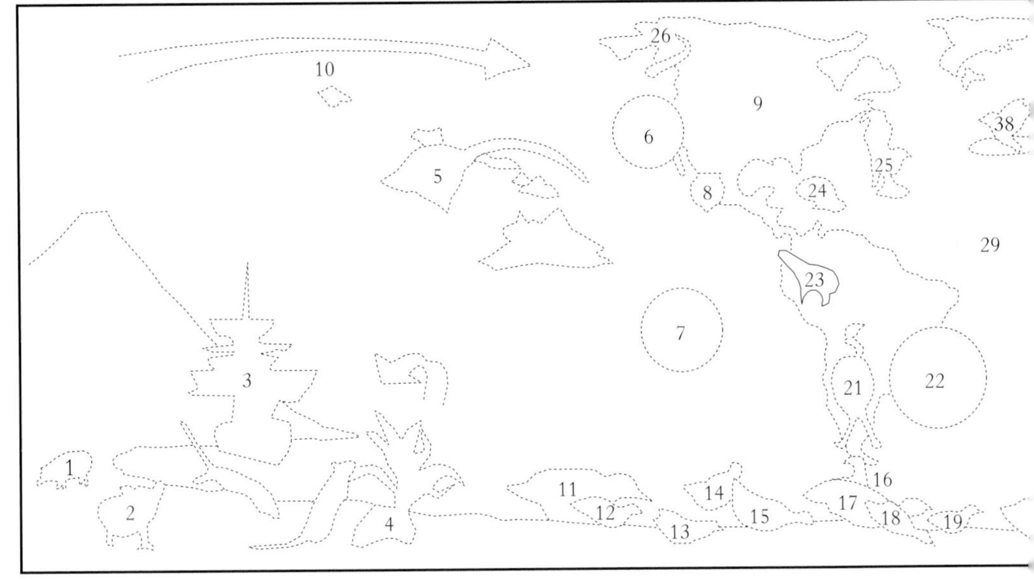


1. Feral pig
2. Norway rats
3. European exploration of remote oceanic islands
4. Coconut palm
5. Hawaiian honeycreeper and mosquito
6. California condor
7. Easter Island
8. Mexican grizzly bear
9. North America
10. Human migration through Beringia
11. Barredwing rail
12. Aukland Island slate-breasted rail
13. Laysan rail
14. Ponape crake
15. Samoan wood rail
16. Chatham Island banded rail
17. Lord Howe wood rail
18. Wake Island rail
19. Hawaiian rail
20. Tahitian rail
21. Rhea
22. Rio de Janeiro
23. Tapir
24. Haitian solenodon
25. Piping plover
26. Alaskan pipeline
27. House sparrow
28. Feral pigeon
29. Caiman hunting
30. North Sea oil industry
31. Gorilla

*A recent history of the world, from an ecological perspective.*

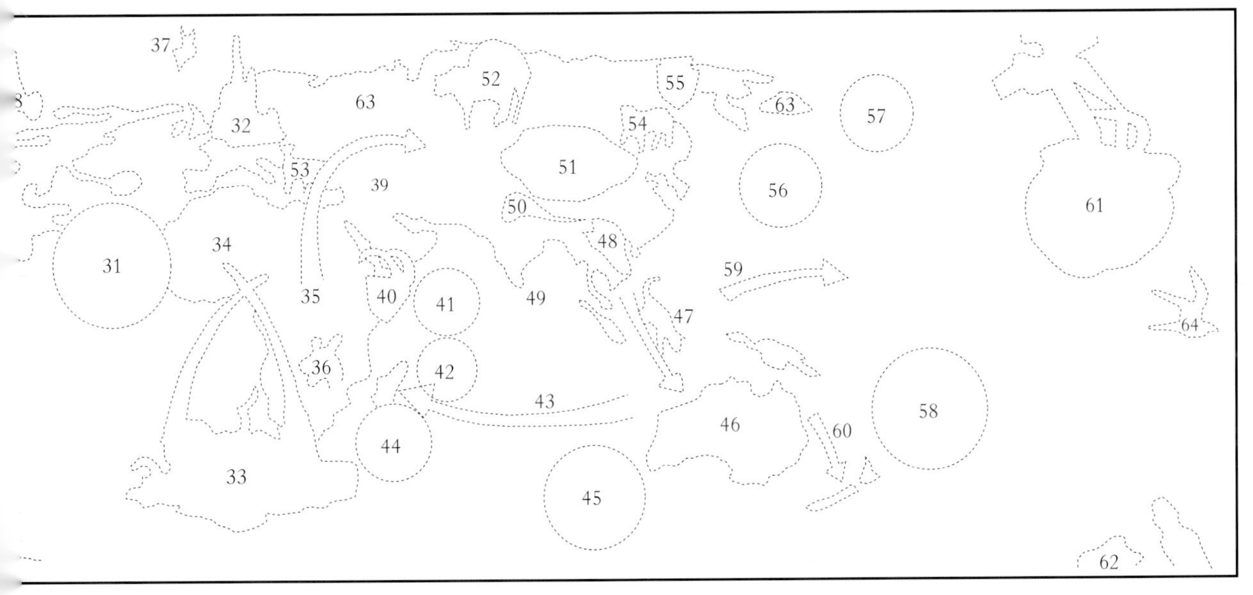


32. EuroDisney
33. Big game hunting
34. The Sahara
35. Origin of modern human
36. Quagga
37. Research whaling
38. Eurasian starling
39. Ostrich
40. Walia ibix
42. *Brookesia peyrieresi*
43. Human migration to Madagascar
44. Dodo and dodo tree
45. Thylacine
46. Feral rabbits
47. Human migration brings dingo to Australia
48. Giant land snail
49. Ceylon elephant
50. Snow leopard
51. Traditional Chinese medicine (bear gall bladder, tiger paw, and rhinoceros horn)
52. Eurasian bison
53. Cretan deer
54. Tiger
55. Kamchatkan bear
56. Shistosoma
57. Hunting woolly mammoth
58. Stephen Island wren
59. Human migration from Asia into the Pacific
60. Human migration to New Zealand
61. Brown tree snake
62. Ocean-borne garbage
63. Stellar's seacow
64. Sandwich tern

then discharged when the ship reaches its destination port. One such invader was the infamous zebra mussel, which made its way into the Great Lakes of North America. The zebra mussel, which originated in Europe, is an extremely efficient filter-feeding organism, straining plankton from the surrounding water so efficiently that it outcompetes native species, which then starve to death. It multiplies rapidly, attaching itself to pipes, boats, and the shells of other mollusks. Zebra mussels clog water intake pipes, thus affecting public water supplies, irrigation, sewage treatment plants, shipping lanes, and recreational activities.

Governments around the world are trying to monitor the alien species arriving in ballast water. A recent study conducted on Japanese ships entering Oregon ports discovered the presence of over 350 alien species being discharged into Oregon waters. Among the most undesirable of such invaders are predatory crabs, which are capable of wreaking havoc on shellfish beds. Such an invasion began in the 1990s with the appearance of the green crab in Washington State. The green crab feeds on small clams, and is capable of decimating local populations of clams and snails.

Plants also suffer a great deal from biological invasions. Because plant seeds are usually small, they are easily transported long distances, and often can colonize and take over new ecosystems quickly. In different areas of the United States, introduced plants make up between 7% and 48% of the total plant diversity. Many of these non-natives, such as kudzu, were deliberately introduced to control soil erosion. Others were introduced as agricultural crops. On rangelands, invasive plants such as cheatgrass crowd out more nutritious native plants, cause soil erosion, and pose threats to native wildlife.

Even more deleterious than these plant invasions has been the transport of plant pathogens from one part of the world to another. Dutch elm disease decimated elm trees in both England and the United States after it was accidentally imported. The introduction of the chestnut blight from Asia to America in 1890 drove the American chestnut tree to the verge of extinction in less than fifty years. In Australia, native Jarrah forests have been destroyed due to the introduction of a root fungus imported from Eastern Australia.

## Winners and Losers

Predicting winners and losers in the future can be as perilous for biologists as it is for stockbrokers. In both cases, however, there are some clear signs of what may prosper (and even diversify) and what may die out.

One clear insight into predicting whether a species will flourish or not comes from the size of its geographic range. In the late 1980s, biologists J. Brown and M. Maurer showed that species of North American birds with small geographic ranges almost always had low population densities within those ranges. In other words, there are virtually no species of birds in North America that are both narrowly distributed and abundant within their small geographic ranges. On the other hand, birds with widespread geographic ranges are usually abundant in most regions within their ranges. Although we see this all around us, it is not an obviously intuitive finding; but it is a generalization that has the utmost importance for picking winners as well as losers among species in the coming years.

The correlation between range size and abundance can be understood by looking at the geometry of species ranges. Geographic ranges exist because they encapsulate all the areas where a species can exist. Thus the outer limits of a geographic range tend to be less favorable areas for the species than the interior of the range. If the size of the less favorable perimeter is large relative to total area of the range, it becomes limiting to the overall population, and small geographic ranges have higher perimeter-to-area ratios than large ones. Not surprisingly, then, when a large geographic range is suddenly broken up into many smaller ranges, the abundance of the species will drop. Geographic range fragmentation can thus influence the likelihood of extinction by affecting the rate of extinction of local populations that find themselves confined to "habitat islands."

This correlation between range size and abundance is the single greatest nightmare of conservation planners. The urbanization and "agriculturalization" of the world has been fragmenting the ranges of most wild species while greatly expanding the geographic ranges of agricultural species. This effect will in essence spell doom for a majority of the world's rare species, many of which are "megamammals." Once again, the conclusion of the Age of Megamammals is the functional reuniting of Gondwanaland.

*Climate is one of the most difficult things to predict. But one thing is certain: man's effects on it have been enormous and the future will be problematic.*

FIVE

# THE NEAR FUTURE
## A New World

There may not be room enough on Earth for both animals and human development.

—ROGER DISILVESTRO, *The Endangered Kingdom*

Can we predict what the future course of evolution may be? It is sometimes tempting to make fanciful conjectures about the nature of future species, but it is also generally nonscientific. Trying to predict the shapes, colors, and appearances of new species would be fantasy, not science. Yet it *is* possible to make other types of predictions, based on what we know through the study of the evolutionary record.

The first thing we can be sure of is that following the current mass extinction there will be empty ecological niches, and these niches will be filled by newly evolved species. But which species will fill a given niche—here is where a crystal ball is necessary. Stephen Jay Gould has long argued that chance will be the major arbiter in deciding which species will replace a newly extinct taxon. For example, perhaps the extinction of rhinos and elephants will trigger the evolution of some group of antelope species toward gigantism to fill the gap, or perhaps the replacement will come from domestic horses—which it will be is mostly a matter of chance. Yet other evolutionists are not so sure that Gould is correct in this view. Paleontologist Michael McKinney (among others) has argued that the best chances of filling the new niches belong to what he calls *supertaxa,* species belonging to groups that are themselves composed of numerous species. Examples of such groups are the rodents, snakes, and passerine birds—all of which are extremely species-rich. McKinney pointed out that since these groups are generally composed of generalists rather than specialists, their members are abundant—and that the same traits promoting numerical dominance also lead to an ability to diversify rapidly over long periods of time. Another characteristic of this group is small body size.

Second, predicting the makeup of any future biota requires an understanding of what the new range of habitats on Earth will be. While the emergence of humanity as the dominant species on Earth has changed things such as the degree of gene flow between once isolated populations and the commonness of alien invasions, perhaps the biggest change has been in the nature of habitats. Humanity has transformed the Earth's surface by producing physical habitats that have never existed before. Through the emergence of megacities, the changeover from old-growth to managed agricultural forests, the spread of agricultural landscapes, the fragmentation of native landscapes by roads, changes in the ecology of the oceans due to the disappearance of large fish, mangrove, coral reefs, and seagrass beds, and the chemical makeover of land and water habitats with pesticides and other chemical pollutants, humans will undoubtedly have a marked effect on future evolution. Natural selection will produce new varieties of life to deal with a set of new environmental conditions never before encountered on the planet.

## Cities

In the late 1970s I flew from the Yucatán Peninsula to Los Angeles, with a stopover in Mexico City. While Los Angeles was well known to me at that time, I had never been to Mexico City, and looked forward to the experience.

Our plane lifted from the lushly verdant Yucatán on a luminous day, and we flew over a starkly visible Mexico. The flight was not very long, and a vista of mountains and forests passed far beneath us. Eventually I spotted a distant mountain, larger than the others, and as we approached I was filled with wonder. Never had I seen a mountain like this before, perfectly dome-shaped, brown in color, impossibly tall, a vision that enlarged and degenerated into implausibility. Our pilot headed straight toward the summit of this great mount, and just as we were about to crash into it, I realized what it was: the air over Mexico City, a mountain of pollution covering the huge sprawl below. Even in the 1970s Mexico City may have been the world's largest city, and it was a sure vision of the future, of what too many poor humans in one place can produce in what we now call "megacities."

Very few animals transform their habitat as extensively as *Homo sapiens* does, and of all our transformations, perhaps none are so visible as the formation of cities. While many of humanity's changes, such as deforestation and the planting and maintenance of agricultural fields perturb and then change one type of biological system into another, the building of cities is a widespread transformation of the organic into the largely inorganic. Termite nests, prairie dog towns, and a few other examples are slight intimations of this process, but true concrete jungles have of course been altogether unknown prior to ours.

Humans had been building up to large-scale cites for millennia, but the advent of the Industrial Revolution changed the nature of cities forever. Once places where trade was centralized and people lived, cities became, during the nineteenth century and throughout the twentieth century, the places where factories and industries were located. The effect was to bring pollution into the bedroom. Ebenezer Howard, an early urban planner, described these new cites as "ill ventilated, unplanned, unwieldy, and unhealthy." The French architect Le Corbusier was more poetic in his denunciation: "They are ineffectual, they use up our bodies, they thwart our souls."

Urbanization is clearly transforming the Earth. At the dawn of the twenty-first century over half of humanity lives in urban areas. By 2030, demographers estimate that twice as many people will live in urban areas as in rural regions. As the twentieth century comes to a close, cities and urban regions occupy about 2% of the Earth's land surface. Yet they use some 75% of the Earth's resources, and release a concomitant amount of waste material. They are not only centers of human population, but centers of human production and consumption.

Urbanization has five major effects: it produces an increased demand for natural resources in the area surrounding the city; it obliterates the natural hydrological cycle at the site of the city; it reduces biomass and alters species composition in and around the city; it produces waste products in high concentrations that can alter the environment around the city, and it creates new but altered landforms through landfilling and reclamation. What is "reclaimed," of course, is generally natural wetlands or lakes. Cities replace natural forests, grasslands, and other vegetation with concrete and brick. These changes vastly affect the flow of water through the site of the city, generally causing water movement to accelerate.

Canadian economist William Rees has brought attention to the concept of the "footprint" of cities—the area of land required to supply them with food and timber products, as well as the area (and plant growth) necessary to absorb the carbon dioxide output they generate. One such "footprint," for London, was calculated by Herbert Girardet of Middlesex University in Great Britain. London was the first of the "megacities," being the first urban area to attain a population of a million people. Using the analyses pioneered by Rees, Girardet calculated that the footprint size of London is 125 times its surface area. The city covers 159,000 hectares, and its footprint is thus 20 million hectares. This is larger than all of the productive agricultural land in Great Britain put together—for just 12% of Great Britain's total population.

All of the key activities of cities—transport, heating, manufacturing, the generation of electricity, and the provision of services—rely on a steady and regular supply of fossil fuels. London requires 20 million tons of petroleum each year—and in the process of converting that petroleum to energy, discharges 60 million tons of carbon dioxide into the atmosphere. Every day London disposes of nearly 7,000 tons of household waste.

The urbanization process has been accelerated by economic, political, and biological factors. The liberalization of political systems around the world in the closing decades of the twentieth century resulted in a surge of economic activity that was accompanied by urban growth. Unfortunately, many cities are decaying as they grow, or cannot keep pace with population increases to provide sanitation, food, and chemical-free environments. Thus urban poverty is driving many of the changes found in cities today. Current growth projections suggest that there will be a minimum of 100 "megacities" with populations of over 5 million people by 2030. These environmental anomalies will have profound effects on both local and, ultimately, global climate.

Climate statistics for recent decades have shown that most cites are warmer than the countryside surrounding them. Thus the boundary between the countryside and city forms a steep thermal gradient surrounding an urban heat island. The heating of cities is a product of several factors. The first is the absorption of solar energy. Most roads and many city roofs are made of dark material that absorbs more solar energy than the surrounding countryside. These city surfaces also have a high capacity for heat storage; concrete and tar roof surfaces both store heat by day and release it at night to a far higher degree than does a vegetated land surface. Second, cities are warm because they generate a great deal of artificial heat through their energy output. Finally, the concentration of large numbers of people and machines in cities causes marked changes in air quality. The release of huge amounts of carbon dioxide and other greenhouse gases that may not readily dissipate out of the city center itself provides an insulating blanket around the city core.

One of the most salient effects of cities is their formation and accumulation of waste. In earlier times the solution was simple: dump the garbage in the poorer neighborhoods. While this system has changed in more developed areas, it still goes on in poorer countries. Perhaps the most striking example of this practice was in the Philippines. In the 1950s Manila began to dump its escalating volumes of waste in a particularly poor neighborhood. The mini-mountain of trash was named Mt. Smokey because of the haze from burning methane, and it towered 130 feet above sea level in a city built at sea level. Even more striking than its size, however, was the biomass of humans that this heap of garbage sustained. In the early 1990s it was home to some 20,000 people who made a living out of sifting through and recycling the waste and eating its remains; they lived there, fed there, and bred there.

The vast amount of waste material generated by cities is transforming the Earth, and just as surely promoting new strains of evolutionary development. Open pits and piles of waste are breeding grounds for human pathogens, but even more so, they are habitats for legions of insects, birds, and small mammals living off the abundant foodstuffs. It is estimated that New York City receives 20,000 tons of food each day for its human inhabitants. About half is transformed into human energy; the rest becomes human sewage and "wastage." This vast resource is an evolutionary target of opportunity for the animals that are now exploiting it. It is clear that 10,000 tons of food material a day was not appearing in the small area now known as New York City prior to the presence of the city; its relatively sudden appearance (by the standards of geologic time) is a sure stimulant for exploitation, and might spur evolutionary change as well, so as to optimize that exploitation.

The first consequences of evolutionary change within this system are probably behavioral—the aspect of evolution that is least visible but probably among the fastest to appear through natural selection. The body of a rat or a cockroach is admirably "preadapted" for living in this new habitat, the garbage dump, and may need little morphological, or body plan, adaptation to exploit it. Yet the new challenges of city living have undoubtedly affected the genetically coded behaviors of these animals, and will continue to do so. Since for the most part we view the animals living off our waste and garbage as pests, we do our best to exterminate them. Any behavior that staves off this fate will be selected for and quickly incorporated into the genetic systems of those animals. We may not necessarily see new species

*Some fauna have adapted surprisingly well to inhospitable urban landscapes, and will continue to do so—but they will be fewer and fewer.*

of city-dwelling organisms as the centuries pass (although we may), but there will be consequential evolution taking place nevertheless, much of it behavioral.

Behavior is not all that will be changed; physiological adaptation will certainly be necessary for city inhabitants as well. A consequence of cities and their wastes is the presence of toxins. Even the most expensive methods of waste disposal, such as high-tech incinerators and so-called "sanitary" landfills, generate toxins. The most important among these are heavy metals, chlorine compounds, and dioxin, all found even in incinerator ash. These compounds and many others make their way into groundwater systems and thereby enter the ecosystem of the city. Animals inhabiting cities, and especially those living in areas with high concentrations of toxins, such as sewers, groundwater systems, and at the base of landfills, might undergo adaptation to withstand high levels of otherwise lethal chemicals, high acid or base concentrations, and even the elevated temperatures found in smoldering landfills.

The ultimate changes, however, may be morphological. We may well see the evolution of a bird beak specialized for feeding out of tin cans, or rats developing oily fur to slough off toxic wastewater. Similarly, new breeds of house cats might evolve larger size to deal with more ferocious rats. But might something completely different evolve? Could we see the evolution of an animal specializing in the most obvious resource of all: human beings?

## The Future Global Climate

Let us imagine a world of long ago, of very long ago: the world of 750 million years ago. It is a time when the first animals are just appearing. It is also the time of "snowball Earth."

The discovery that, at several times in its history, the Earth was covered from pole to pole with ice is one of the major geologic finding of the late twentieth century. The just-completed Ice Age pales in comparison to these long-ago times. Ice locked Earth in its grip, covering both land and sea. There was virtually no life on the planet, save for a few oases of warmth next to undersea volcanoes. The discovery that these "snowball Earth" events occurred not just once but repeatedly, albeit long ago, shows but one swing of the messy pendulum we call climate. It is also a lesson in how extreme climate change can be—and perhaps soon will be.

No one disputes that humanity is rapidly changing the composition of the atmosphere, although there is still great debate about whether or not those changes are causing a rise in mean global temperature, also known as global warming.

Anthropogenic, or human-induced, production of gases such as carbon dioxide, methane, chlorofluorocarbons, sulfur dioxide, and nitrogen oxides have been rising dramatically since the Industrial Revolution. All of these gases have the ability to absorb infrared radiation and reradiate it back to Earth, producing the well-known "greenhouse effect." To better understand the conditions that will face life in the future, we must better understand what the gas inventory of the atmosphere will be.

As we all know, predicting the weather is a chancy business. Trying to make valid long-range predictions for the next several days is hard enough. Doing the same for the next few thousands of millennia seems impossible. Yet in some ways the long-term view is clearer than the short-term view. Almost all scientific information to date suggests that global warming will be a long-term reality.

Predictions about the possibly of global warming over the next few decades and centuries come from a class of models known as General Circulation Models (GCMs). A starting point of these models is the prediction that the amount of carbon dioxide in the atmosphere will double over the next century. This doubling is sure to have profound ecological effects, including greater temperature increases in mid-latitude temperate and continental interior regions than across the rest of the globe, decreases in precipitation in these same mid-latitude regions, and an increase in severe storm patterns.

Such changes will affect the entire biosphere, but will have their most marked effects on plant communities. Because there is so much paleontological information about how plant species and communities fared during the rapid climate changes accompanying the end of the Ice Age over the past 18,000 years, there is some room for optimism that reasonable projections about oncoming climate changes can be made.

According to paleobotanists, four prime lessons from the near past are applicable to the near future. First, it seems that species, rather than whole communities, respond to climate change. Over the past 18,000 years, the species compositions of various North American forests have changed considerably, yet the forests themselves have persevered. Whole communities and biomes do not respond to climate change, but instead change their species compositions. Second, the responses of individual species to climate change are often accompanied by a time lag. Especially rapid climate change tends to overwhelm many plant species because they are incapable of dispersing rapidly enough to move with the changes. For example, the eastern hemlock tree can disperse at a rate of 20–25 kilometers per century. However, climate patterns can move at a rate of over 300 kilometers per century.

The net result can be local extinction of a species if climate change is sufficiently fast. The third insight is that patterns of local disturbance will change as climate changes. Fire is one of the principal causes of disturbance in modern forest ecosystems; as climate changes, the pattern and frequency of major forest fires will change as well. Changes in such disturbance patterns may produce a greater change in an ecosystem than the climate change itself. Fourth, it seems that multiple environmental changes can produce unpredictable responses in ecosystems. If enough sources of change come to bear on a given ecosystem, its responses may not be predictable. We may be on the verge of seeing the formation of terrestrial plant communities unlike any that have existed in the past—not through the formation of new species (although that may happen as well), but through novel compositions of groups of species that have no ancient community analogues.

Another factor will be the response of plants to increased levels of carbon dioxide ($CO_2$) in the atmosphere. Many plants increase their photosynthetic activity and growth rates in response to elevated amounts of $CO_2$. A result of increased $CO_2$ levels will therefore be greater global plant productivity, faster growth, and perhaps larger plants of some species. On the other hand, there are distinct differences among plants in their responses to raised levels of $CO_2$. Some plants (using an enzyme system known as the $C_4$ metabolic pathway) are already saturated with $CO_2$ in the present-day atmosphere, and will not respond with faster growth or productivity if $CO_2$ is elevated globally. A second, more common group of plants (those using the so-called $C_3$ metabolic pathway) will respond to the increased $CO_2$ with enhanced growth. There are other determining factors as well. Plants living in high-stress conditions and those from highly disturbed habitats will show little effect, while plant species that are stress-tolerant will do better. Perhaps it will come as no surprise that the winners in a new, $CO_2$-rich environment may be weedy species.

Although $CO_2$ levels will have an important effect on plant community compositions and growth rates, by far the single most significant factor affecting plant community composition and growth is water availability. There is an enormous amount of variation among plant species in their ability to withstand drought, so future patterns of precipitation and runoff around the globe will affect the makeup of plant communities most. As global temperature changes affect water distribution across the planet, plants will be forced to adapt to rapidly changing conditions.

In his chapter "Appreciating the Benefits of Plant Biodiversity," from the final twentieth-century installment of the best-selling series *The State of the World,* botanist John Tuxhill suggests that the first signs of changing carbon dioxide levels are already being observed in tropical rainforests. Tuxhill notes that the "turnover rate" of tropical forests—the rate at which old trees are replaced by younger trees—has been increasingly steadily since the 1950s. As a result, the forests under study are becoming "younger" through increasing domination by shorter-lived trees and woody vines that grow faster than the tall hardwood trees that make up the old climax communities. Such trends will favor a radical changeover in the species composition of the tropical forest. Tuxhill also notes:

> Global trends are shaping a botanical world that is most striking in its greater uniformity. The richly textured mix of native plant communities that evolved over thousands of years is increasingly frayed, replaced by extensive areas under intensive cultivation or heavy grazing, land devoted to settlements or industrial activities, and secondary habitats—shorter lived "weedy," often non-native species.

Under these conditions, can we expect substantial future evolution in forest communities? Since a very high diversity of plants has already evolved, there are probably many species "preadapted" for the new conditions that are being produced now by global atmospheric change. While one can speculate and dream up new plant species evolving to take advantage of higher carbon dioxide levels, the reality is that very little new evolution may occur within the dwindling forests of the planet.

## The Oceans

What separates the current mass extinction from those of the past is that little or no extinction has yet occurred in the oceans, and that changes in temperature, toxicity, and other environmental factors there have been minor compared with those on land. But for how long?

While the oceans have not undergone an equivalent of the megamammal extinction characterizing the last 50,000 years on land, it would be a mistake to assume that some extinction has not occurred. The exploitation of fisheries stocks has not eliminated more than a handful of species, but its effects, from the large-scale disappearance of whales and other marine mammals to the reduction of the large fish species used for human food, have utterly transformed the biological makeup of the oceans and the way in which energy flows through its communities.

Although the great British zoologist Thomas Huxley opined that "all the great sea fisheries are inexhaustible," the results of a century of exploitation contradict that statement. The reduction of large carnivores in the sea represents a radical restructuring of the single largest habitat zone on Earth. Perhaps this restructuring will instigate future evolution, but can any outburst of evolution occur while fishing pressure exists?

Humans depend on the oceans for food, raw materials and minerals, and transportation lanes, and the strains of those uses are showing. It is estimated that the proportion of marine fisheries stocks that are overexploited has climbed from almost none in 1950 to between 35% and 60% as the twentieth century came to a close. The most pressing threats to the oceans, according to the 1,600 scientists contributing to the United Nations 1998 Year of the Oceans program, are species overexploitation, habitat degradation, pollution, climate change, and species introductions. As one of these scientists put it, "Too much is taken from the sea and too much is put into it."

The use of fisheries stocks at the end of the twentieth century was staggering. The world's human population received 6% of its total protein, and 16% of its animal protein, from the sea; and over a billion people relied on fish for at least 30% of their animal protein supply. Up to 90% of this catch comes from coastal zones (which also supply at least 25% of the Earth's primary biological productivity).

The major fish stocks showing a marked decline in catch totals include sharks, tuna, swordfish, salmon, and cod. As these stocks decline, new species are exploited in their place. During the 1980s five low-value species—Peruvian anchovy, South American pilchard, Japanese pilchard, Chilean jack mackerel, and Alaskan pollock—accounted for the majority of new landings. Moreover, the efficiency of fishing is now such that formerly prosperous regions of the sea are becoming biological deserts. The once rich Grand Banks off Canada are now bereft of the cod that were once so abundant; the king crab fishery in Alaska has collapsed; the orange roughy fishery of the South Pacific is essentially nonexistent. Trawling—the practice of dragging nets and chains across the ocean bottom—is now so widespread that it is estimated that every bit of the world's continental shelves is dragged at least once every two years.

The rising human population is also affecting coastal zones. Two-thirds of the world's largest cites are located on seacoasts, and the environmental effects of these burgeoning human populations are radically changing the seas. The destruction of seagrass beds and mangrove regions to allow human settlement has had a marked effect on fish stocks, since these habitats are breeding grounds for many important species.

The seas are also the final resting places of most anthropogenic pollution. River systems dump waterborne waste into the sea; winds carry airborne pollution into the sea; excessive nutrients from nearby cities create dense carpets of algae that ultimately rot in the sea. Such algal blooms take oxygen out of the water and create large "dead zones." Much of the Gulf of Mexico is now afflicted by such dead zones. Other consequences of nutrient loading in the oceans are an increase in red tides and an increase in paralytic shellfish poisoning. Synthetic organic chemicals also end up in the sea, as do radioactive materials and heavy metals such as mercury.

All of these factors make the oceans one of the most potent cauldrons for future evolutionary change. Yet of all of the ecosystems on Earth, the oceans may see the *least* amount of species-level extinction and, perhaps paradoxically, the *most* new evolution. The reasons are several. First, despite our best efforts, even completely overexploited and subsequently "crashed" fisheries stocks do not go extinct. But they do not recover as long as fishing continues—and as long as there is a large human population, there will be overfishing. At the same time, the vast size of the oceans and their lack of native habitability by humans will always provide a buffer for marine creatures. Try as we will to invade them, the oceans will always be far less perturbed than the land. Thus, as new species are formed (beginning, always, as tiny isolated populations), there is less chance that human intervention will immediately stop the new speciation process.

Second, the removal of top carnivores—the species most exploited by humans—will leave a void that will be filled by natural selection and new speciation. Although humans exploit the upper parts of marine food webs, the lower trophic levels are barely touched. Humans do not exploit, for example, the copepods, small worms, and other invertebrates making up the majority of the ocean biomass. New species will evolve to fill the vacuum created by the drastic reduction in numbers of commercial fish stocks.

What will evolve to take the place of the larger fish species? Because fish, according to the fossil record, appear capable of evolving rapidly, it may be that the new species will be other fish. But if large fish species evolve to replace those reduced or rendered extinct by overfishing, the same trend may happen again, and they will become the overexploited. It is more likely that either many small fish species will evolve, or the positions in the upper parts of the marine food chain will be filled by larger invertebrates.

## Agricultural Fields

The largest single habitat type on the surface of the land will soon be agricultural fields. Most of the ancient forests and the drier grasslands and savannas of the Earth have been, or are in the process of being, converted into farms, and this conversion will be a major contributor to new evolutionary events. But if farmers' fields predominate, a second major habitat type increasing in size will be deserts. Quite often, fields turn into deserts through poor agricultural practices and reductions in water availability.

By the late twentieth century the option of expanding grain production by cultivating more land had virtually disappeared. From 1950 to 2000, increases in the harvest of grain came from the conversion of forest and native grassland into grain-producing fields, but this option has been exhausted. The few regions left to exploit include the cerrado of Brazil, a semiarid rangeland in the east-central part of the country, the area around the Congo River in Africa, and the outer islands of Indonesia. At the same time, vast areas currently used for growing grain will be lost to human housing, or through soil erosion and land degradation. The amount of cropland per person on Earth is expected to decline from 0.23 hectares in 1950 to 0.12 hectares in 2000, and to 0.07 hectares by 2050. The area of cropland in India, for example, will not rise, but it will have an estimated additional 600 million people to feed by 2050. By the same time China will need to feed a total of 1.5 billion people.

The winners in the agricultural environment will be insects, rodents, and predators on both. As in the case of domesticated animals, it is likely that a great deal of evolutionary change has already occurred since the inception of agriculture nearly ten thousand years ago, unremarked by early humans. A taxonomist assessing the insect and rodent makeup of the world prior to the start of human agriculture might be surprised at how many species that are common today did not exist then. Rodents are known to have some of the fastest evolutionary rates on Earth; a thousand years is more than sufficient time to create new species, and the ten thousand years since agriculture began may have seen a vast proliferation of small animals living among the crop rows. The same process has surely occurred among insects, perhaps on even a vaster scale than among the rodents. Because animals of this size are not readily observed or perturbed by human mitigation efforts, the surge of evolution is likely to continue. Armies of new ant, beetle, and rodent species seem a probability.

## The Human Population

Ten thousand years ago, there may have been at most 2 to 3 million humans on Earth. There were no cities, no great population centers; humans were rare beasts, scattered in nomadic clans or groups, or at best in settlements of little lasting construction. There were fewer people on the entire globe than are now found in virtually any large American city. By two thousand years ago that number had swelled almost a hundredfold, to 130 million or perhaps as many as 200 million people. The billion mark was reached in the year 1800, and there were 2 billion people by 1930, 2.5 billion in 1950, 5.7 billion in 1995, and approximately 6.5 billion in 2000. At this rate of growth, the human population is expected to exceed 10 billion sometime between 2050 and 2100, assuming an annual increase of 1.6%. While this rate is somewhat reduced from the 2.1% growth rate characterizing the 1960s, it remains a staggering figure.

In 1992 the United Nations published a landmark study calculating potential human population trends, which arrived at several estimates. By 2150, the human population could reach about 12 billion, if human fertility figures fall from their present-day levels of 3.3 children per woman to 2.5 children. If, however, the faster-growing regions of the world continue to increase in population and maintain their current fertility levels, average fertility worldwide will increase to 5.7 children per woman, and the human population could exceed 100 billion people sometime between 2100 and 2200. The latter figure seems beyond the carrying capacity of the planet. Officially, the United Nations uses three estimates for the year 2150: a low estimate of 4.3 billion, a medium estimate of 11.5 billion, and a high estimate of 28 billion.

Predicting future population numbers is a difficult endeavor because of the many variables involved. The definitive work in recent times is Joel Cohen's 1995 *How Many People Can the Earth Support?* Cohen's conclusions are stark:

> The possibility must be considered seriously that the Earth has reached, or will reach within half a century, the maximum number the Earth can support in modes of life that we and our children and their children will choose to want. . . . Efforts to satisfy human wants require time, and the time required may be longer than the finite time available to individuals. There is a race between the complexity of the problems that are generated by increasing human numbers and the ability of humans to comprehend and solve those problems.

There are, of course, many reasons why the higher figures may not be reached. Human disease, such as HIV or some other pathogen, may affect these figures; famine or war could also markedly reduce them. Barring such calamities, our population of approximately 6 billion humans at the turn of the millennium will at least double in slightly more than a century to a century and a half. Once this figure of approximately 12 billion humans is reached, it is assumed that the population will stabilize.

More than 200 years ago, the British scientist Thomas Malthus described the single most intractable problem with human population growth. While our population numbers increase exponentially, human food supply tends to increase on a linear scale as more land is devoted to agriculture. The inescapable conclusion is that the human population will tend to outgrow its food supply. In a related fashion, the human population is likely to outstrip its supply of untainted and unpolluted fresh water.

Water may indeed be the most critical factor in determining the maximum human population that the Earth can support. While the Earth's stock of water is immense, most of it is salt water held by the oceans. The amount of fresh water is far less—only a small percentage of the total. Moreover, about 69% of that fresh water is locked in glaciers, permanent snow cover, or aquifers more than a kilometer deep, all inaccessible to humans. About 30% is present as accessible groundwater, leaving 0.3% in freshwater lakes and rivers. This totals about 93,000 cubic kilometers of fresh standing water on the Earth's surface. This water does not stay in place, however: it evaporates into the atmosphere or sinks into groundwater stocks. Thus, a total of between 9,000 and 14,000 cubic kilometers of renewable fresh water is available for human agriculture each year.

Humans use water for more than agriculture. People drink about 2 liters of water per day in temperate climates, and perhaps three times this amount in arid climates. But drinking is the least of human water consumption. In a developing country all household uses—including cooking, consumption, and washing—amount to about 7 to 15 cubic meters of water per year per person. The average person in a developed country uses twice this amount. Yet these figures pale when the amount of water needed to feed each person on Earth is calculated. It takes approximately 200 tons of water per year per person to raise sufficient wheat to maintain that person on a "model-skinny" (a.k.a. starvation) diet. This translates to about 350 to 400 cubic meters of water per person per year—a whopping 300 gallons per day. Eating meat requires even more water. If 20% of the diet comes from animal (meat and dairy) products, about 550 cubic meters of water per person per year are

*Some fauna have adapted to a more aquatic though degraded setting, as in this new freshwater "tree of life."*

required, whereas the typical American diet requires 1,000 cubic meters of water each year to produce.

Water specialists have calculated maximum, median, and minimum estimates for the amount of water available to agriculture: the maximum figure is 41,000 cubic kilometers of water, the median 14,000, and the minimum 9,000. Assuming that all of that water is used for human food production and consumption, the high figure would sustain a global population of between 25 and 35 billion people on the American diet, and between 100 and 140 billion at near-starvation levels. Yet assuming that all the available water can be used for agriculture is ridiculous; about 80% is actually used for other purposes, especially industrial uses. A more reasonable estimate is that 20% of the total water volume is available for agriculture. With this figure, the world can support at most between 5 and 7 billion people on the American diet, assuming that 41,000 cubic kilometers of water are available, and only 1.1 to 1.6 billion assuming the lowest figure of 9,000 cubic kilometers. Even on a starvation diet, the world can support between 20 and 30 billion people assuming that the maximum projection of water availability is correct, but only between 4 and 6 billion if the minimum figure is correct. Thus the current human population of the Earth may already exceed its carrying capacity based simply on water availability.

## Parasites and the New Manna

Seldom has the world seen a more striking transformation: in little more than three hundred years, the majority of the biomass found in terrestrial animals has shifted from many species to only a few. The most prominent of these new winners are humans and the domesticated animals bred to feed them. Since evolutionary forces tend to respond to the presence of new resources, we might expect prodigious amounts of new evolution among human, cow, sheep, and pig parasites.

The evolution of parasites is usually in lockstep with the evolution of new host species. Parasites require particular adaptations for the host bodies they inhabit. Humanity has long been present on Earth, and we have long had our share of parasites. But the huge population increase since the end of the eighteenth century has created large human populations in regions where few people once existed, especially in the more humid and torrid tropical regions, and a consequence of this change has surely been natural selection for more, and more efficient, human parasites. While the same trend should be creating new human predators as well as human parasites, the successful evolution of an efficient new human predator is a long-term, and ultimately futile, process: as soon as we humans get wind of any evolutionary change putting ourselves, and especially our children, in harm's way, we will institute immediate and surely successful eradication efforts. Killing new parasites, however, is a far more difficult endeavor, especially those of very small size, such as microbial forms. It is in this arena that some of the most interesting and fecund new species of the coming biota may be found.

## Toxicity

The agencies of humankind are rapidly changing the chemical makeup of the surface of the Earth and its waters and oceans. Most of this chemical change comes from human-induced pollution, the result of municipal, industrial, and agricultural wastes. These wastes can be specifically categorized as nutrients, metals, and synthetic and industrial organic pollutants. All of these pollutants pose challenges to living organisms, and it is certain that future evolution will in some cases be triggered by reaction and adaptation to new levels of these substances.

Nutrients such as nitrates and phosphates are the cause of eutrophication, an explosion of biological activity. Anthropogenic sources of these substances include synthetic fertilizers, sewage, and animal wastes from feedlots. Much of this nutrient pollution makes its way into rivers. A. Goudie and H. Viles, in their book *The Earth Transformed,* suggest that nitrate and phosphate levels in English rivers have increased between 50% and 400% in the last 25 years alone.

Metals, the second major class of pollutants, occur naturally in soil and water, but their natural concentrations have been vastly increased by human activity. The most toxic to humans are lead, mercury, arsenic, and cadmium. Other metals are poisonous to marine organisms, including copper, silver, selenium, zinc, and chromium.

Since the 1960s, synthetic and industrial pollutants also have been manufactured and released into the environment in large quantities. The synthetic organic compounds currently released into the environment now number in the tens of thousands, and many are hazardous to both terrestrial and aquatic life at even low concentrations. The most dangerous include chlorinated hydrocarbon insecticides (such as DDT), PCBs, phthalates (which are used in the production of polyvinyl chloride resins), PAHs, which result from the incomplete burning of fossil fuels, and DBPs, which are disinfectants. All of these compounds mimic naturally occurring organic compounds and are readily absorbed by living organisms, producing birth defects, genetic abnormalities, health problems, and death. In humans some of these compounds are implicated in lowered sperm counts.

A further chemical change is brought about by acidification. The well-known phenomenon of acid rain is changing the pH of many terrestrial environments, causing biological problems and even local extinctions among some organisms.

## Ozone Depletion

In the 1980s, scientists, and soon thereafter the general public, became aware of the loss of ozone from the upper regions of the atmosphere. Much of the thinning of the ozone layer has been caused by the release of chlorofluorocarbons (CFCs) into the atmosphere. A long-term reduction of ozone would have profound evolutionary effects, as ozone screens out ultraviolet radiation, which is poisonous to living matter given sufficient exposure. If long-term ozone depletion continues, organisms will be forced to evolve new structures or physiological pathways to deal with excess UV radiation.

All of these accumulating chemical changes will require specific physiological adaptations in a host of organisms. Although the most obvious effects of evolution are visible changes in body types, far more evolutionary change takes place at the behavioral and physiological levels. In these cases, evolutionary change is not readily apparent. Yet, in the increasingly toxic environments of Earth, they will remain the most common types of future evolution.

## Prophecy

The factors described above can be used to pick the potential evolutionary "winners" of the future: organisms adapted to cities or agricultural fields and capable of living in polluted water or air. Much future evolution may be invisible, taking place among already existing animals through changes in behavior and physiology. Can some vision of our world, even a millennium from now, be imagined? With apologies to H. G. Wells (and to those who require that books about science remain "serious" and dry), here is mine.

The Chronic Argonaut smiled briefly, closing his well-thumbed novel. He pushed the lever forward and sped into the future. At the year 3000 A.D. he came to a stop. His time machine was located on a small grassy field in northwestern Washington State. In the distance the familiar Cascade Mountains looked just as they had when he had last seen them, on the first day of the year 2000. A thin rain was falling, not unusual for Seattle at this time of year, no matter what the century, he thought. But it was a warm rain, and he noticed how tropical the air felt. He began to stroll.

The park was filled with plants, and at first he took no notice of them. But with wonder he began to notice the large leaves and brilliant colors of foliage he had never seen in this area before. Citrus trees were visible, and acacias, and as he looked at the greenery around him he was struck by the lushness and clearly tropical nature of the vegetation. Nearby he could see buildings, clearly different in composition and architecture, but recognizable nevertheless. He was a bit crestfallen. Other than the dramatic changes in vegetation and climate, he found that the future was not so very different.

He came upon roads, and people. They looked perfectly ordinary, a mixture of the races of his own time familiar and present still. But the streets were crowded. To his surprise and wonder, the University of Washington was still present, a maze of buildings now completely covering the once parklike and open campus. With the friendly help of students he made his way to the library, and found what he was looking for: an encyclopedia for the year 3000. The news within was not good.

The human population had stabilized at 11 billion. The total number of species on the planet was still unknown, but the list of the large animals that had gone extinct since his own time was explicit. Africa was especially hard hit. Gone were the African wild ass, mountain zebra, warthog, bushpig, Eurasian wild pig, giant forest hog, common hippopotamus, giraffe, okapi, Barbary red deer, water chevrotain, giant eland, bongo, kudu, mountain nyala, bushbuck, addax, gemsbok, roan antelope, waterbuck, kob, puku, reedbuck, hartebeest, blue wildebeest, dama gazelle, sand gazelle, red-fronted gazelle, springbok, suni, oribi, duiker, ibex, Barbary sheep, black-backed jackal, wild dog, Cape otter, honey badger, African civet, brown hyena, aardwolf, cheetah, leopard, caracal, aardvark, pangolin, chimpanzee, red colobus, and guenon. Also extinct were the indri, black lemur, and aye-aye in Madagascar. Also gone were the pygmy chimpanzee, mountain gorilla, brown hyena, black rhinoceros, white rhinoceros, pygmy hippopotamus, scimitar-horned oryx, white-tailed gnu, slender-horned gazelle, and Abyssinian ibex.

In Asia the list contained the giant panda, clouded leopard, snow leopard, Asiatic lion, tiger, Asiatic wild ass, Indian rhinoceros, Javan rhinoceros, Sumatran rhinoceros, wild camel, Persian fallow deer, thamin, Formosan sika, Père David's deer, Malayan tapir, tamaraw, wild yak, takin, banteng, Nilgiri tahr, markhor, lion-tailed macaque, orangutan, Indus dolphin, and douc langur. In Australia the victims included the Parma wallaby, bridled nailtail wallaby, yellow-footed rock wallaby, Eastern native cat, numbat, hairy-nosed wombat, and koala. In North and South America the list included the spectacled bear, ocelot, jaguar, maned wolf, giant otter, black-footed ferret, giant anteater, giant armadillo, vicuña, Cuban solenodon, mountain tapir, golden lion tamarin, red uakari, and woolly spider monkey. All had been either endangered or threatened in his own time. None had been saved from extinction—not with 11 billion human mouths to feed, year in, year out.

There was other news as well. The sea level had risen by 15 feet, drowning many of the world's most productive land areas and requiring humanity to turn most of the larger forest areas into fields. India, China, and Indonesia were the world's most populous countries, and all had become heavily industrialized. World temperatures had risen sharply as coal replaced oil as the chief energy source for the planet. But for him the saddest news was of the coral reefs. Like the rainforests, they were now restricted to small patches of territory amid the huge range they had once dominated.

He was able to glean some information about the fate of his own species. Computers, robots, and nanotechnology had radically changed human professions. But there was still an enormous gap between wealthy and poor nations. While there had been innumerable wars and skirmishes (some of which had been going on even in his own time), two larger events had completely altered the human psyche. Both involved outer space. While the early part of the second millennium had seen an ongoing effort to explore outer space, the energy behind that effort seemed to be dissipating. Humankind had reached Mars with manned missions, and had even mounted a manned mission to Europa, the distant moon of Jupiter. To the delight of astrobiologists, life—true alien life—had been found in both places. But that life was microbial. Nothing more complex than a bacterium seemed to exist elsewhere in the solar system. The material cost of these two visits had been staggering, and despite the discovery that there was indeed life in space beyond the Earth, no practical reason for returning could be found. There were no great mineral deposits or other economic reasons for this type of space flight, and certainly no reason to colonize either of these otherwise inherently hostile worlds. It proved to be far more cost-effective to colonize, and essentially "terraform," Antarctica than it was to carry out the same endeavor on Mars. Although space flight into low Earth orbit and occasional visits to the moon to maintain the manned astronomical observatories on its far side continued, no further expeditions to the far reaches of the solar system could be justified.

The second disappointment lay in the stars. Even with the great advances of technology during the second millennium, they were no closer at the end of that millennium than at the beginning. There was no great breakthrough in propulsion systems that might allow speeds approaching anything near the speed of light; the vision of faster-than-light starships, or travel through wormholes, remained the domain of moviemakers. Nor was there any further stimulus to visit the stars, since in spite of a millennium of searching, no signals from extraterrestrial civilizations had ever been received. SETI, the search for extraterrestrial intelligence, maintained its lonely vigil through the centuries, but to no avail. The stars remained distant and mute. Humankind looked wistfully into a closed sky, and then gradually gave up looking. There would be no escape to the stars. There would be no zoo of new extraterrestrial animals to assuage the guilt and longing of the human race in a new world largely bereft of large animals. All scientific results suggested that while microbes might be present throughout the galaxy, animals would be rare. Humans lived on a Rare Earth.

He left the university, looking for other changes. He made his way downtown through the sparkling city. As a child he had loved to go to the fish market, a place where the entire panoply of edible marine biodiversity was always on cheerful display: the many varieties of salmon, the bountiful rockfish, lingcod, black cod, sole, halibut, steelhead, sturgeon, true cod, hake, sea perch, king crabs, Dungeness crabs, rock crabs, box crabs, oysters, mussels, butter clams, razor clams, geoducks, horse clams, Manila clams, octopus, squid, rock scallops, bay scallops, shrimp, deepwater prawns. All this sealife came from just the cool waters of his home state. But the market was gone, and in none of the food stores he visited could he find any seafood at all for sale. There was chicken, beef, pork, mutton, and lamb, and there were many varieties of vegetables, many new to him. But no seafood, no food at all from creatures not cultivated or domesticated but harvested from the wild. Nothing.

He walked through the city, now so ancient, at least in human terms. There were no songbirds. But there were crows by the thousands.

He looked for new varieties of things. But the birds, the squirrels, the domesticated dogs and cats—and the people—all looked the same.

A thousand years had not yet brought about a new fauna growing from the ashes of the old. That would require more time. But then again, that was something that the Time Voyager—and his species—had in almost limitless supply. All the time in the world.

*The transformation of domestic plants and animals—from age-old selective breeding to today's genetic engineering—has had an enormous impact on all living things on the planet.*

SIX

# THE FIRST 10 MILLION YEARS

## The Recovery Fauna

One of the most remarkable features in our domesticated races is that we see in them adaptation, not indeed to the animal's or plant's own good, but to man's use or fancy.

—CHARLES DARWIN, *On the Origin of Species*

Sitting high atop a hill with an unbroken view to the west, Seattle's Harborview Hospital enjoys a commanding vista of the vast city sprawling along the shores of the inland waterway known as Puget Sound. On clear days the rugged Olympic Mountains float above the far western horizon, seemingly trying to snare the setting sun with their jagged rocky tops, while to the south Mt. Rainier looms like some upstart Baskin Robbins flavor of the month. Perhaps no other public hospital in the world enjoys such a magnificent setting.

Such bucolic pleasures are largely lost on Harborview's clientele, however. Most who come here are beyond caring about such things, for those passing through these doors usually do so only as a last resort. And they do not come alone. The transients, drug addicts, gunshot victims, and uninsured who make up a large percentage of Harborview's patients often bring with them exotic new strains of microbes, organisms that are assuredly only recently evolved.

Beginning in the middle part of the twentieth century, scientists and doctors waged a campaign of eradication against bacterial illnesses, using the then newly developed antibiotic drugs. The result was a mass extinction of bacteria—a concentrated interval of death resulting in the loss of uncountable individual microbes, and for all intents and purposes, the extinction of whole species. Smallpox, rabies, typhoid, rubella, cholera: the ancient microbial scourges of humankind were wiped out. The bacteria causing these ancient plagues were faced with two alternatives: evolve or die. Most died. But a few evolved forms resistant to antibiotic drugs. Fifty years after their invention, these "miracle" drugs have unleashed a diversity of new drug-resistant species that would never have evolved but for the influence of humanity. These are surely but the beginning of a host of new microbial species.

And so too with the biosphere, except that the "antibiotic" is us. As a result of our antibiotic influence, and the current pulse of extinction it has engendered, many currently living species will die. Some, however, will survive and thrive, becoming the rootstock of a new biota. Some have already done so, for one of the precepts of this book is that significant portions of the "recovery fauna" that follows any mass extinction are already with us, and dominating terrestrial habitats, in the form of domesticated plants and animals. Evolution obviously continues, but much of it is now "directed" for human purposes, or occurs as a by-product of human activities.

Charles Darwin began his *On the Origin of Species* with a chapter on domestication. Before introducing any other data or argument, he pointed to the many varieties of domesticated animals and plants as one of the clearest proofs of the existence of organic evolution—in this case, the evolution of new types of animals and plants bred to serve as food or as companions to humanity.

As with most of his conclusions, Darwin was right about this point. But we can take it one step further. Domesticated animals and plants are the dominant members of what can be called the "recovery fauna" accruing, directly or indirectly, from the extinction of megamammals (and that extinction's impetus toward the development of agriculture). That many of these animals are taking the functional place of extinct or endangered megamammal species is no accident. The cows, pigs, sheep, horses, and other familiar domesticated animals now covering the world's grasslands rapidly replaced the many species of extinct or endangered large wild grazers. The stimulus to that evolutionary changeover has, of course, been the stern hand of humanity.

## Characteristics of Domestication

Although most animal species can be "tamed," or to some extent habituated to the presence of humans if raised by them from a young age, domestication goes far beyond this simple behavior modification. Domesticating a species requires not only a concerted effort over extended periods of time, but certain pre-existing features of the species in question. In the past this effort was made only for reasons such as enhanced food yield, transportation, or protection from predators. Domesticated animals are the evolutionary results of human-induced "unnatural selection."

Very few large mammals have been domesticated. Biologist Jared Diamond of UCLA showed that of the 150 species of terrestrial, noncarnivorous mammals larger than about 30 kilograms, only 14 have been domesticated. All but one of these originated in the Eurasian region; the only New World exception to this rule is the llama. All of the domesticated species are derived from wild species with similar characteristics: all seem to show rapid growth to maturity, an ability to breed in captivity, little tendency to panic when startled, an amenable and tractable disposition, and a social structure and hierarchy that permits domestication. All of these characteristics were further selected for by a brutal form of "natural selection": those individuals that showed the favored characters were allowed to breed; those that did not were killed.

Interestingly, neurologist Terry Deacon has noted a further characteristic: all domesticated animals appear to have undergone a loss of intelligence compared with their wild ancestors. This observation makes one wonder whether humans, compared with their ancestors, have also been "domesticated," and undergone a similar reduction in intelligence.

The first domesticated species may well have been the dog. All modern dog species seem to be derived from the Asiatic wolf. Although the first anatomically

*Foremost among the survivors of the next millennium will be those species that humans have had a hand in developing: domesticated plants and animals.*

modern dogs may date back as far as 12,000 years ago, bones of canids recognizable as "dogs" first appear around human campsites between 7000 and 6000 B.C. Even though wolves can still interbreed with dogs, implying that true biological speciation has not occurred, behavioral differences make such interbreeding rare; thus the modern dog varieties as we known them are functionally distinct from wolves. The domestication of the dog also led to distinct anatomical changes. Compared with wolves, dogs have smaller skulls, shorter jaws, smaller teeth, and pronounced variation in coat color. Dogs also appear to be less intelligent than wolves. Most dog varieties now recognizable were produced in the eighteenth and nineteenth centuries; prior to that, dogs generally were used for hunting (hounds) or tending flocks (sheepdogs).

The bones of domesticated livestock first appear in the fossil record slightly later than those of dogs. Sheep and goats came first; the earliest evidence for their domestication, dated at around 8000 B.C., comes from various sites in southwestern Asia (the areas now composed of Israel, Iran, Jordan, and Syria). Cattle were derived from an entirely extinct species of wild cowlike creatures. Domesticated pigs also date back to about 8000 B.C. Four thousand years later, the domesticated horse was developed from wild horses in Eastern Europe. (The ancestor of the domestic horse, called Przewalski's horse, still exists, in small numbers, in reserves in Poland.) Donkeys, water buffaloes, and llamas were domesticated at about the same time, while chickens and camels were not brought into the menagerie until about 2500 B.C. A wide variety of smaller "pets" have been domesticated as well: house cats, guinea pigs, rabbits, white rats, hamsters, and various birds. All are the result of human effort.

In almost every case, the transformation of a wild species into a domestic species involves substantial physical and behavioral modification. It has long been thought that this process occurred in stages, beginning with "taming" and progressing to gradual genetic transformation. How such great genetic change was forced in such a short period of time has always been a puzzle. New research by geneticists may have provided an answer. There appears to be a master gene controlling a complex of genes that in turn affect tameness, reaction to stress, coat color, facial morphology, and social interactions. By changing a single gene rather than an entire complex, domestication could proceed relatively quickly.

## Domesticated Crop Species

While it may be argued that the near coincidence of the beginning of agriculture and the end of the Age of Megamammals is just that—coincidence—a strong argument also can be made for cause and effect. The extinction of many of the larger animals upon which humans depended for food, coming as it did with a major climate change (affecting plant and small animal resources also used for food) may not have been mere chance.

While cereals such as wild wheat and barley were harvested as much as 12,000 years ago, it seems that the first domestication of plants took place about 10,000 years ago, at the time when the last mammoths, mastodons, and many other larger animal species were dying out in North America and had just disappeared in Europe and Asia. This was the time when food-gathering peoples began to collect the seeds of wild plants and replant them in the ground. The domestication process appears to have involved the natural hybridization of several wild species, followed by selection by humans for desired characteristics. Thus "domestication" of plants, like that of animals, involved the genetic modification of the wild species through a very rough form of natural selection: those plants with usable traits were kept; those without were killed. Since the trend in plant modification has been toward an increase in the size of the edible or usable parts, most plant species have lost the ability to disperse widely, and protective mechanisms such as thorns have generally been lost as well.

The number of domesticated plant species is relatively quite small. There are more than two hundred thousand species of angiosperms, or flowering plants, yet only ten of these provide the vast majority of human food. Among these ten are grasses and cereals such as wheat, rice, and maize, which are all characterized by seeds rich in starch and protein. Cereals are planted on 70% of the world's cropland and produce about 50% of the calories consumed by humanity. Other plants in the top ten include sugarcane, yams, potatoes, bananas, soybeans, and manioc. Worldwide, about three thousand species of plants are used as human food, but only about two hundred of these have become domesticated.

## The Transgenic Revolution: Building Weeds

The genetic engineering our ancestors used to introduce new characters into their agricultural crops and domestic animals was crude but effective: save the favored varieties and let them breed; kill off the others. But in the twentieth century a new type of genetic engineering appeared—one that alters the genomes themselves. This new way of introducing novelty is sweeping the agricultural regions of the Earth, and will surely have unintended consequences. It may be that the transgenic revolution will bring novelty into the biotic world in ways almost unimaginable—and not all of them desirable. It appears, for example, to be on the verge of creating "superweeds."

Modern genetic technology allows the transfer of genetic material from one species to another. This new genetic information is permanently integrated into the genome of the second species, conferring new traits upon it. For all intents and purposes, a new type of organism is let loose into the biosphere each time this is done. The organism thus transformed is called a *transgenic* plant, animal, or microbe. These transgenic creatures have not arisen through the natural processes of evolution, but they are among the most portentous developments for the future of evolution on this planet.

Transgenic organisms are possible because of the existence of certain genes capable of "jumping" form one chromosome to another. The first discovery of jumping genes was made by American geneticist Barbara McClintock in the 1940s. McClintock was studying the genetics of maize (corn), and observed that certain genes, such as those responsible for seed color, were capable of moving from one chromosome to another. The significance of this discovery was largely overlooked until the 1970s, when it was independently rediscovered by other researchers examining the ways in which certain bacteria develop resistance to antibiotics. The genes, or sections of DNA, coding for these specific characters in bacteria do not actually "jump"; instead, they produce copies of themselves, which are inserted at other points either on chromosomes or in genetic code–carrying organelles called plasmids.

The discovery of these jumping genes, technically called transposons, unleashed a torrent of research in the 1980s and into the 1990s. These peculiar strings of DNA have the ability to repeatedly cut and paste themselves into different parts of an organism's genetic code. What made them famous—and may eventually make them infamous—is that the transposons of one organism can be used to paste new genetic information into the DNA of entirely unrelated organisms.

Much of the research using transposons was conducted on fruit flies. The fruit fly *Drosophila* is one of the stalwarts of experimental genetics, since it breeds quickly and its genetic code is well known. In the early 1980s Gerald Rubin and Alan Spradling of the Carnegie Institute discovered a transposon in *Drosophila* that could be used to incorporate new genetic information into these files. They succeeded in changing the genetic code of the transposon and re-inserting it into the fly. With this operation, they had succeeded in creating a fly with a new genetic code that could be passed on to a succeeding generation. They had created a transgenic species—one entirely new to the world, a species with a genetic code created not by nature, but by science.

These early experiments altered very little in the new fly. The majority of its genetic code was the same as that of the unaltered species. But certain characters, such as color or eye type, could be changed. Further work showed that certain fruit fly transposons were not only useful in changing the genetic code in fruit flies, but could be placed into entirely unrelated species. A method had been developed that allowed true genetic engineering of insects.

The goals of genetic alteration of insects are laudable. Insects create havoc in human society in two ways: they serve as vectors of diseases (e.g., malaria, yellow fever, and some types of sleeping sickness), and they consume a large proportion of human crops. Genetic engineering is attempting to mitigate both of these problems. Nevertheless, the results of this work have been slow. In certain disease-spreading mosquitoes, geneticists have so far been able to change eye color, but they have yet to alter the structures involved in spreading disease-causing microbes. To further aid this process, geneticists have infected the target insects with viruses that act like transposons. The virus, once in the body of the insect host, can alter the way a disease is passed on. Some crop pests, such as the Mediterranean fruit fly and the screwworm, have been successfully targeted using transgenic or other genetic techniques (e.g., producing sterile members of a species that spread among the viable members of the population).

While transgenic techniques are just coming into use in controlling insect pests, such tools are already used widely in crop plants. Genetic engineering has succeeded in adding new genes to the DNA of various crops, whereas conventional breeding only adds variants to an already existing genetic complex. So, while domestication has enhanced the valued characteristics of many plant species, transgenic research has added entirely new characteristics, such as greater tolerance of heat and drought, greater resistance to insect predation and disease, and greater yield.

The genetic engineering techniques used in agriculture are widespread and sophisticated. Genetic engineers can move genes from virtually any biological source into crop species. Genes have been added to engineered crops from organisms as diverse as chickens, hamsters, fireflies, and fish, as well as from a slew of plant and microbial species. The new transgenic plants are genuinely novel organisms, some of them containing the genetic codes of plants, animals, and microbes in a single species.

The addition of new genes to various plant species has yielded spectacular dividends in terms of crop yield. But this new technology also poses substantial risks and clearly will affect the future of evolution on this planet. The creation of new plant types could affect the biosphere in various ways. The most important of these is the possibility that newly inserted genes will jump to other, nonengineered species (such as weeds) or move out of the agricultural fields altogether into wild native plant populations. It is this potential intermixing of new genes with those of already established plant species beyond the boundaries envisioned or designed by agricultural scientists that could have the most interesting—and potentially biosphere-altering—effects. Under rare circumstances, if new traits of transgenic species escape into the wild, weeds adaptively superior to native plants could be created. Since most of the traits being transferred into transgenic crops, such as hardiness, resistance to pests, and growth rate, increase their fitness relative to the original species, there is great potential for transgenics to become, or to help produce, new weedy species.

There are several avenues by which the genes of transgenic crops can become established in the wild. First and simplest, the transgenic crop itself can escape and become a weed. Second, the transgenic crop can release pollen into the wild, which can be incorporated into a wild relative of the original transgenic host. The incorporation of the new genes into the wild plant creates a new transgenic weed.

Weeds have many definitions, which are often colored by human values. In agriculture, they are plants that occur in the wrong place at the wrong time (some plants are weeds in certain situations and favored crops in others—lawn grass, for example). A more human-specific definition of a weed is any plant that is objectionable to or interferes with the activities or welfare of humans. Nevertheless, weedy species have a number of characteristics:

Their seeds germinate in many environments

Their seeds remain viable for long periods of time

They grow rapidly

Their pollen is usually carried by nonspecialized pollinators or by the wind

They produce large numbers of seeds

They produce seeds under a wide range of environmental circumstances

They usually show vigorous vegetative reproduction or regeneration from fragments

They often compete either by choking out other plants or by producing toxic chemicals deleterious to other plants

From this list, we can see that the characters of weeds are also highly desirable in crop plants. One goal of transgenic technology, therefore, has been to confer characteristics of weeds on crop species. Transgenes spliced into crops may alter traits such as seed germination ability, seed dormancy, or tolerance of either biotic or abiotic factors such as pests, drought, heat, or disease, creating a more persistent or resistant species in the process. Such new traits may enhance the new crop's ability to invade other habitats. Genes affecting seedling growth rates, root growth rates, and drought tolerance are currently being developed.

Jane Rissler and Margaret Mellon of the Union of Concerned Scientists have studied the ecological risks of transgenic crops in thorough detail. One of their most important concerns relates to the transformation of nonweedy crop species into weeds through genetic engineering. They note that a widely held notion is that changing a non-weed into a weed involves the conversion of many genes, not just the two or three currently used in transgenic crops. Changing corn from a crop to a weed, for example, would involve a number of genetic changes, since corn is one of the most human-dependent (and thus intolerant) plant species on Earth. Other crops, however, already possess many weedy traits, and thus one to three new traits could indeed create a new weedy species. Examples include alfalfa and other forages, barley, lettuce, rice, blackberries, radishes, raspberries, and sunflowers.

Transgenic species may produce secondary effects as well. The invasion of transgenic plants into new habitats affects not only the invaded plant populations, but the entire ecosystem, including the suite of animals living within that ecosystem. Perhaps more dangerous than the conversion of crop plants into weeds is the escape and transference of new genes into already existing weeds, making them "superweeds." The transfer of disease resistance or pest resistance to established weedy species may change a familiar weed into an even more formidable pest. Weeds that have developed resistance to herbicides because of escaped or transferred transgenes are already appearing in some parts of the world.

Agribusiness, which counts transgenic technology as the jewel in its technological crown, is in the business of feeding the world's humans—and making profits in doing so, of course. One of the great fears of those producing transgenic crops is that farmers will simply take the seeds from the first crop and use them ever after, and will not need to buy new seeds from the corporation that produced them in the first place. Like the software industry, which fears the copying of its products above all else, the major biotechnology companies dealing in transgenic crops have been searching for some way to stem the illegal use of their products after the first purchase. The solution is something known as the "terminator gene."

The first terminator gene was produced by the large American biotechnology firm Monsanto, and was engineered to protect Monsanto's patent rights on several types of transgenic crops. It is a genetic modification that prevents seeds from germinating after the season in which they were sold. In the works are genes that will allow but a single crop and not produce future seeds—a bit like the seedless watermelon, but more efficient.

The great fear is that such terminator genes will jump to unmodified varieties of crop plants. If the terminator gene in a tomato plant jumped to other varieties of tomatoes, there a is real potential of plants never producing the crop they were intended to produce.

## Does Evolution Have a Future?

Our species has learned how to circumvent the normal rules of evolutionary change: we have learned how to build new species. Have we also achieved the ability to alter those rules? Norman Myers of Oxford asks this question in his prescient and disturbing 1998 paper, "The Biodiversity Crisis and the Future of Evolution." Myers makes a subtle but important point: humans pose "pronounced threats to certain basic processes of evolution such as natural selection, speciation, and origination." Myers has cried wolf before, but the wolf was always there, dining happily on flocks of the world's species. Is he being alarmist in this case? Although many have warned of a biodiversity crisis, Myers alone warns of an *evolution crisis.* He bases his conclusion on two perceptions: first, that we have entered a new phase of mass extinction, and second, that the normal recovery period from mass extinction will not pertain to this one; in fact, the recovery will be considerably delayed.

Myers cites three aspects of this particular mass extinction that will affect its evolutionary outcome (and which make it different from any mass extinction of the past):

> Its onset was extremely fast (compared with those of the past), within a single century or two, and thus there will be scant opportunity for ecosystem reorganization and evolutionary response.
>
> There is currently a higher biodiversity on the planet than at any time in the geologic past, so that if 50% of species are lost, the total number going extinct will be higher than in any mass extinction of the past.
>
> During past mass extinctions, plant species have been largely spared, but that may not be the case in the current mass extinction.

The current mass extinction may be unique not only in what it kills, but in how its recovery proceeds. In past times the tropical regions of the world have served as storehouses for recovery. Because they have always held the greatest diversity of species on the planet, they have long served as "evolutionary powerhouses"—areas that seem to spawn new species and new types of species at a higher rate than other parts of the world. Paleontologist David Jablonski of the University of Chicago has shown that innovation can be related to geography. Innovation in evolution is the appearance of evolutionary novelty, and the tropical regions seem to be home to more innovation than other regions. Yet the tropics are now the sites of the highest densities of humans and the greatest human population increases. This pattern may curtail the evolution not only of new species, but also of new types of species.

The current crisis in biodiversity may also substantially reduce the number of new species evolving a large body size. Megamammals need very large habitat areas to survive; it may also be true that they need equally large areas to speciate. With the reduction of wild habitat, and especially free rangeland, virtually everywhere on Earth, there may be no way for large mammals and other vertebrates to produce new species. Therefore, a consequence of human population growth and habitat disturbance may be not only the extinction of large mammals, reptiles, and birds, but the inability of new large species to take their place, simply because the mechanism of speciation for large body sizes has been derailed by environmental fragmentation.

## Implications for Conservation Planners

A large and vibrant community of conservationists, scientists, politicians, and laypeople are actively engaged in intensive efforts to preserve biodiversity. One of the most important such efforts is habitat preservation. Yet even the most Herculean of efforts will save only patches of habitat in a sea of agricultural fields and spreading human landscapes. As long as humanity rules, it is doubtful that hundreds of thousands of miles of unfenced, unimpeded native habitat will be available to replace the species already lost since the end of the Ice Age. This fact has led Norman Myers to pose the following questions:

> Is it satisfactory to safeguard as much of the planetary stock of species as possible, or should greater attention be paid to safeguarding *evolutionary* processes at risk? This is an entirely new way of looking at the world—not in terms of losing species, but in terms of losing pathways of speciation. Perhaps the motto should be "save speciation" rather then "save species."
>
> Of prime importance is the question of biodisparity—the number of body types. There could be many species on Earth, but few body types. Is it enough to save a large number of species if we fail to save biodisparity as well?
>
> Should the evolutionary "status quo" (the current makeup of the Earth's biota) be maintained by preserving precise phenotypes of particular species that will enable evolutionary adaptations to persist, thereby leading to new species? For example, should two elephant species be maintained, or should we keep the option of elephant-like species in the distant future?
>
> Is there some minimum number of individuals necessary not just for the survival of a species, but the survival of the potential for future evolution in that species? Should the slow breeders (the megamammals) be given greater attention than, say, the rapidly breeding insects? Are we in a triage situation?
>
> How do we assess the relative importance of endemic taxa as compared with evolutionary fronts such as origination centers and radiation lineages? Myers thinks it far more appropriate to safeguard the potential for origination and radiation than any individual species. Let endemic taxa go.

This last recommendation is heresy by the rules of modern conservation. It has long been argued that endemic centers—those regions that contain species found nowhere else—are among the most important places to save. But Myers's point is that endemic centers exist because they have not produced large numbers of successful species. Endemic centers are often living museums of ancient species that do not have much potential for future evolution.

## The Weeds of Humanity

The vast human enterprise has created a new recovery fauna, and will continue to provide opportunities for new types of species that possess weedy qualities and have the ability to exploit the new anthropogenic world. Chief among these will be those species best preadapted for dealing with humanity: flies, rats, raccoons, house cats, coyotes, fleas, ticks, crows, pigeons, starlings, English sparrows, and intestinal parasites, among others. These and our domesticated vassals will dominate the recovery fauna. Among plants, the equivalents will be the weeds. According to many seers, this group of new flora and fauna will be with us for an extended period of time—a time span measured in the millions of years. And if humanity continues to exist and thrive (as I believe it will), this recovery biota may dominate any new age of organisms on Earth.

A sense of how long the recovery fauna may last was estimated in a disturbing paper published in the prestigious journal *Nature* in the spring of 2000. The authors, James Kirchner and Ann Weil, posed a question: how quickly does biodiversity rebound after a mass extinction? How long will the world exist at a very low biodiversity? The answer, it turned out, was far longer than anyone had heretofore estimated. By analyzing the fossil record of all recoverable organisms (compiled by the late Jack Sepkoski of the University of Chicago), Kirchner and Weil found that fully 10 million years passed by, on average, before the biodiversity of the world recovered to its pre-extinction values. Even more surprising than this long lag period between extinction and full recovery was their finding that it occurred whether the extinction was small or large. We paleontologists had assumed that the time to recovery would somehow correlate with the magnitude of the extinction—that after a small extinction, the biosphere would recover quickly, and that it was only after the greatest of the mass extinctions that a long recovery period was necessary. But to the surprise of us all, Kirchner and Weil found this not to be the case—10 million years was necessary even after the smaller extinctions. They concluded their paper with the following passage:

> Our results suggest that there are intrinsic "speed limits" that regulate recovery from small extinctions as well as large ones. Thus, today's anthropogenic extinctions are likely to have long lasting effects, whether or not they are comparable in scope to the major mass extinctions. Even if *Homo sapiens* survives several million more years, it is unlikely that any of our species will see biodiversity recover from today's extinctions.

It appears that our return to a new biota will take a long time after the mass extinction is finished. And what might that new fauna and flora look like? Some predictions can be made—and such predictions are the subject of the next chapter.

*A poster child for extinction: the thylacine, a prehistoric doglike marsupial once found across much of Australia and Tasmania, was both actively hunted by humans and a victim of competition from human-introduced wild dogs. The last thylacine died in a zoo in 1936.*

SEVEN

# AFTER THE RECOVERY

## A New Age?

Prophets who take themselves too seriously end up preaching to an audience of one.

—GEORGE DYSON, *Darwin among the Machines*

There is an often-articulated notion that if there is any consolation in the prospect—or process—of mass extinction, it is that at the end of the tunnel a new fauna emerges. According to this line of reasoning, the great sacrifice of species is a cleansing of the planet, making way for a renewal. The hope is that after the mass extinction is over, a new Age of some sort will dawn—a better, more diverse Age. It is the parable of the Flood: let us call it "Noah and the Recovery Fauna." After all, this seems to have been the pattern after the two greatest of all mass extinctions, when the dinosaurs took over from the mammal-like reptiles at the end of the Permian, and the mammals from the dinosaurs at the end of the Cretaceous. Could it be that after the current mass extinction, the one group of chordates still waiting for its own "Age"—the birds—will dominate? Will there now be an "Age of Birds," a world of land-based bird herbivores and carnivores, burrowers and climbers, as well as the numerous (or even more numerous) flying forms that characterize this class today? Or might some completely unforeseen group take over, such as giant insects (biomechanically impossible), or something totally new? Unless some altogether new class of vertebrates suddenly appears (highly unlikely), only the birds have yet to hold the honorific of "ruling" the planet. Perhaps the best bet is indeed an Age of Birds. In such a world mammals would still be present, even if they are no longer the evolutionary dominants.

In discussions about the impending biodiversity crisis, this new fauna argument is sometimes used as a rationalization, even a justification. The Age of Mammals—and the Age (or even *existence*) of Humanity—would never have occurred but for the extinction of the dinosaurs, and in like fashion—or so the argument goes—the modern extinction will yield some new age of organisms, perhaps with some new form of global intelligence.

What might this new evolutionary biota be like? Why not something entirely new? Can we imagine an entirely new type of animal that could replace the current evolutionary dominants, the large mammals? This new class would have to have evolved from some currently existing creature, but it could have characteristics and a body plan vastly different from those of the preceding dominants. Such a new body type could exploit some entirely new food type or habitat. Let us imagine such a breakthrough—the conquest of the lower atmosphere by floating organisms called Zeppelinoids.

After the extinction of most mammals (and humanity), Zeppelinoids evolve (let's say from some species of toad, whose large gullet can swell outward and become a large gasbag). The great breakthrough comes when the toad evolves a biological mechanism inducing electrolysis of hydrogen from water. Gradually the toad evolves a way to store this light gas in its gullet, thus producing a gasbag. Sooner or later small toads are floating off into the sky for short hops (but longer hops than their ancestors were used to). More refinement and a set of wings give a modicum of directionality. Legs become tentacles, trailing down from the now thoroughly flight-adapted creatures, which can no longer be called toads: they have evolved a new body plan establishing them as a new class of vertebrates, the Class Zeppelinoida. Like so many newly evolving creatures, the Zeps rapidly increase in size: when small they are sitting ducks (flying toads?) for faster-flying predatory birds. Because their gasbag is not size-limiting, they soon become large. Eventually they are the largest animals ever to have evolved on Earth, so large that terrestrial and avian predators no longer threaten them, reaching dimensions greater than the blue whale. Their only threat comes from lighting strikes, which result in spectacular, fatal explosions visible for miles. The Zeps can never get around this inherent flaw, for there is no biological means of producing the inflammable, inert gas helium and thus avoiding instant death from lightning. But then, life is never perfect, and the Zeps still do well, especially in areas with little lightning.

Now the dominant animals of the world, the Zeps float above the ground like great overgrown jellyfish, snagging with their dragging tentacles the few species of deer (and other herbivorous vertebrates) still extant and stuffing them into a Jabba-the-Hutt-sized mouth. Because the Zeps evolved from amphibians and are still cold-blooded, they have a very low metabolic rate, and thus need to feed only sparingly. Their design is so successful that they quickly diverge into many different types. Soon herbivorous forms are common, floating above the forests, eating the tops of trees, while others evolve into zep-eating Zeps. Still others become like whales, sieving insects out of the skies; in so doing they soon drive many bird species to extinction. The world changes as more and more Zeps prowl the air, floating serenely above the treetops, filling the skies with their numbers, their shadows dominating the landscape. It is the Age of Zeppelinoids.

A fairy tale—but there is a glimmer of reality in this fable. Evolution in the past has produced vast numbers of new species following some new morphological breakthrough that allows some lucky winner to colonize a previously unexploited habitat. The first flying organisms, the first swimming organisms, the first floating organisms, all followed these breakthroughs with huge numbers of new species quickly radiating from the ancestral body type, all improving some aspects of design or changing styles to allow variations on the original theme.

But is the fundamental assumption underlying this scenario—a long period of extinction followed by the emergence of a new class of evolutionary dominants—at all likely? No. For just as humanity has changed the "rules" of evolution that have operated on this planet for hundreds of millions of years, so too has the usual sequence of events following mass extinction been modified.

## Potential Winners of the Future

Picking the evolutionary winners of the future—those species that will evolve to take the place of the "losers" (those going extinct)—is something like trying to pick winners in the stock market, or forecasting the weather. There are some data available to help us make educated guesses, yet the system is so large, and subject to such a plethora of stochastic and chaotic effects, that prediction of specifics is

*Whatever happens to life on Earth, one thing is certain: evolution will not stop. Here is one possible scenario for the evolution of the rat.*

impossible. The colors, habits, and shapes of the newly evolved fauna can only be guessed at. There is information available that might shed light on the future winners, however, in the fossil record.

One of the interesting (and rather unexpected) findings of paleontological research is that higher taxa (the taxonomic categories above genus and species, such as families, orders, classes, and phyla) seem to show typical rates of evolution. The rate of evolution for a taxon can be described in two ways: as the rate at which some morphological character changes through time, or as the longevity of an average species in geologic time. Related to the rate of evolution are origination and extinction rates. Some groups of organisms seem to produce many new species, others very few. And of the species produced, those of some groups last for long periods of time, while those of other groups die out more rapidly.

The importance of understanding evolutionary rates was first pointed out by George Gaylord Simpson, a pioneering evolutionist. More recently, Steven Stanley of Johns Hopkins University has taken up many of the themes of research pioneered by Simpson and added fascinating new insights. Stanley's landmark

1979 book *Macroevolution* explored these themes in detail. Paleontologists know well the groups that show high origination and extinction rates, for these are the most important fossils used in biostratigraphy, the science of subdividing and dating sedimentary rocks using fossils. Good biostratigraphic markers are those fossils that have a short temporal duration—and thus occur in only a few strata—yet are at the same time widespread, common, and have sufficiently distinct morphological attributes that evolutionary change and new speciation events are immediately apparent. Examples include trilobites, ammonites, and mammals, among others. Other groups—sometimes called "living fossils"—show the opposite trends: they speciate slowly, and once originated, rarely go extinct. They are thus useless for biostratigraphy, but fascinating evolutionarily—what is it about such organisms that bestows the equivalent of near-immortality?

A useful way of quantifying evolutionary rates is by arriving at an estimate of *doubling time,* the average amount of time for a particular higher taxon to double the number of species within it. Mammals, for instance, show a doubling time of

3.15 million years. In contrast, bivalve mollusks take 11 million years to double their number of species. Mammals show rapid evolution and bivalves very slow evolution. Within both of these groups there is much variation, including subgroups of rapidly evolving bivalves and slowly evolving mammals. In general, however, it is clear that mammals evolve faster (and produce more taxa in an equal amount of time) than do bivalve mollusks.

There is a third group of taxa that can also be recognized in the evolutionary world. Stanley has proposed the term *supertaxa* for groups of organisms that show both high origination rates (they produce many species) and low extinction rates (their species last a long time). Such groups have a tendency to diversify rapidly, and in so doing they become prime candidates for refilling the world with new species following any mass extinction—including the current one.

The title of champion supertaxon in the world today belongs to the family Colubridae: the snakes. Stanley has suggested that rather than being in an Age of Mammals, we are really in an Age of Snakes! And as the future of evolution unfolds, we may find ourselves in a world filled with many new species of snakes. Another champion "evolver" is the group containing rats and mice, which, perhaps not coincidentally, are one of the prime food sources of snakes. This is probably not what most of us have in mind when contemplating some future world: a world of snakes and rats in untold varieties of form, color, and habit. Joining them will be other rapidly evolving species, many of which can be classified as "weeds" in that they are capable of rapid and wide distribution and are widely tolerant of harsh conditions. Many insect species both evolve rapidly and are consummate weeds (look at all of the flies in the world). Birds are also relatively fast evolvers. Each of these groups can be projected to be very common and proliferate a diversity of new species. Other mammals evolve slightly more slowly than these groups; in general, the larger the animal, the slower its evolutionary rate or doubling time.

Let us imagine some of these outcomes. Snakes could move into niches that they are rare in, or do not completely occupy, today. Many new species of marine snakes seem possible, as do snakes replacing the many small to medium-sized mammalian carnivores now dwindling in numbers. As agricultural fields and cities continue to enlarge in size over the millennia, and even tens of millennia, rodents will proliferate in a variety of new species to take advantage of these new feeding opportunities, and this too will provoke further evolution of new snake species.

Birds and insects are also potential winners. The many species of birds doing well now in urban and agricultural environments could become the rootstock of

*A possible future cladogram, or evolutionary family tree, for the dandelion (bottom to top): original dandelion, cactuslike, aquatic, arboreal, carnivorous, epyphite.*

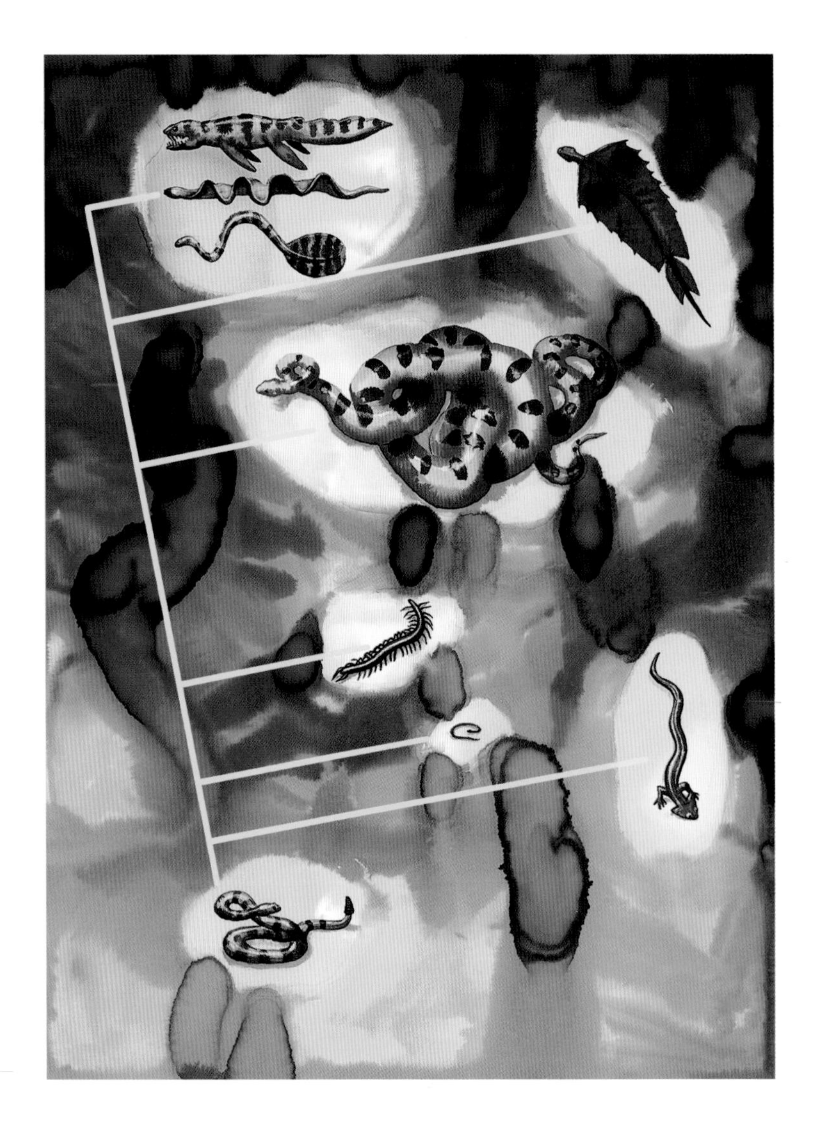

*One future cladogram for the snake (bottom to top): timber rattler, walking, millipede, pygmy giant, flying, three swimming types.*

*One future cladogram for the crow (bottom to top): crow, vulture, shoe bill, raptor, honeyeating, wading, ratite crow.*

many new species. Successful forms such as crows, pigeons, and sparrows might undergo great evolutionary diversification. Of all the birds, crows seem best adapted for coexisting with humanity, and might be among the most successful at diversifying into new species—the dominant species of the new recovery fauna.

The groups mentioned above are all familiar. What about totally new types of creatures, like the fanciful Zeps—the kind that would make entertaining television or great fantasy books? Can there be a reasonable expectation of totally new types of creatures, with novel body plans?

## The Cambrian Explosion and the Expectation of Novelty

The history of life, like any history, has occurred as a vector of time. And as in any history, there is never any going back, at least in any meaningful way. Events and their history create irrevocable changes that make each slice of time unique as it passes through the sequence from future to present to past. In the context of the future of evolution, it appears that there will never again be an Age of Fishes, or Reptiles, or Mammals even approximately similar to those that have occurred in our planet's past. This is a point that conservationists refuse to accept: the Age of Megamammals is over. There will never again be an African veldt with the rich assemblage of mammals now confined to Africa's game parks, and soon enough there may be no game parks at all in Africa. Even if we could somehow remove all humans from the planet in an instant, it is doubtful if things would return to the state they were in 50,000 years ago, at the onset of the end of the Age of Megamammals.

But leaving aside a return to any past era, if humanity suddenly were removed from the planet, could we expect to see new body plans? The reality is that there has been little true evolutionary novelty since the Cambrian Period, 500 million years ago. Although the conquest of land allowed vertebrates and arthropods—the two most successful terrestrial phyla—to evolve and explore new themes of shape, these were only modifications of existing body plans, and even that evolutionary adventure seems far nearer its end than its beginning. The birds are the last class of vertebrates to have evolved, and they did so almost 200 million years ago. Yet there seems to be an expectation that something altogether new will arise. Part of this expectation is raised by what did happen long ago in the past, when evolutionary novelty was cheap, during an event called the Cambrian Explosion.

*One future cladogram for the pig (bottom to top): pig, genetically engineered, rhino pig, aquatic pig, pygmy, giraffelike, garbage-eating.*

For the first 3.5 billion years of its existence, our planet was without animal life, and it was without animals large enough to leave a visible fossil record for another half billion years after that. But when, 550 million years ago, animals finally burst onto the scene in the oceans, they did so with a figurative bang in a relatively sudden event known as the Cambrian Explosion. Over a relatively short time, all of the animal phyla (the large categories of animal life characterized by unique body plans, such as arthropods, mollusks, and chordates) that exist today either evolved or first appeared in the fossil record. Uncontested fossils of animals have never been found in sedimentary strata more than 600 million years old, no matter *where* on Earth we look. Yet the fossils of animals are both diverse and abundant in 500-million-year-old rocks, and they include representatives of the majority of the animal phyla still found on Earth. It appears that in a time interval lasting perhaps 20 million years or less, our planet went from a place devoid of animals that could be seen with the naked eye to a planet teeming with invertebrate marine life rivaling almost any species on Earth today in size.

The rates of evolutionary innovation and new species formation during the Cambrian Explosion have never been equaled. It produced both huge numbers of new species and body plans of complete novelty. That all of the animal phyla would appear in one single, short burst of diversification is not an obviously predictable outcome of evolution. From this observation comes the second finding concerning the Cambrian Explosion that is equally puzzling, if far less well known: The Cambrian Explosion marked not only the start, but also the *end* of evolutionary innovation at the phylum level. Since the Cambrian, *not a single new phylum has evolved.* The extraordinary fact is that the evolution of new animal body plans started and ended during the Cambrian Period.

The lack of new phyla and the paucity of new classes after the end of the Cambrian Explosion may simply be an artifact of the fossil record; perhaps many new higher taxa *did* evolve, and subsequently went extinct. This seems unlikely. It is far more likely that the great surge of innovation marking the Cambrian came to an end as most ecological niches became occupied by the legions of newly evolved marine invertebrates.

Yet there remains a puzzling mystery: why is it that no new phyla evolved after the two great mass extinctions, the Permo-Triassic and Cretaceous-Tertiary disasters? While the Permian mass extinction may have caused the number of species to plummet to levels as low as those found early in the Cambrian, the subsequent diversification in the Mesozoic involved the formation of many new species, but

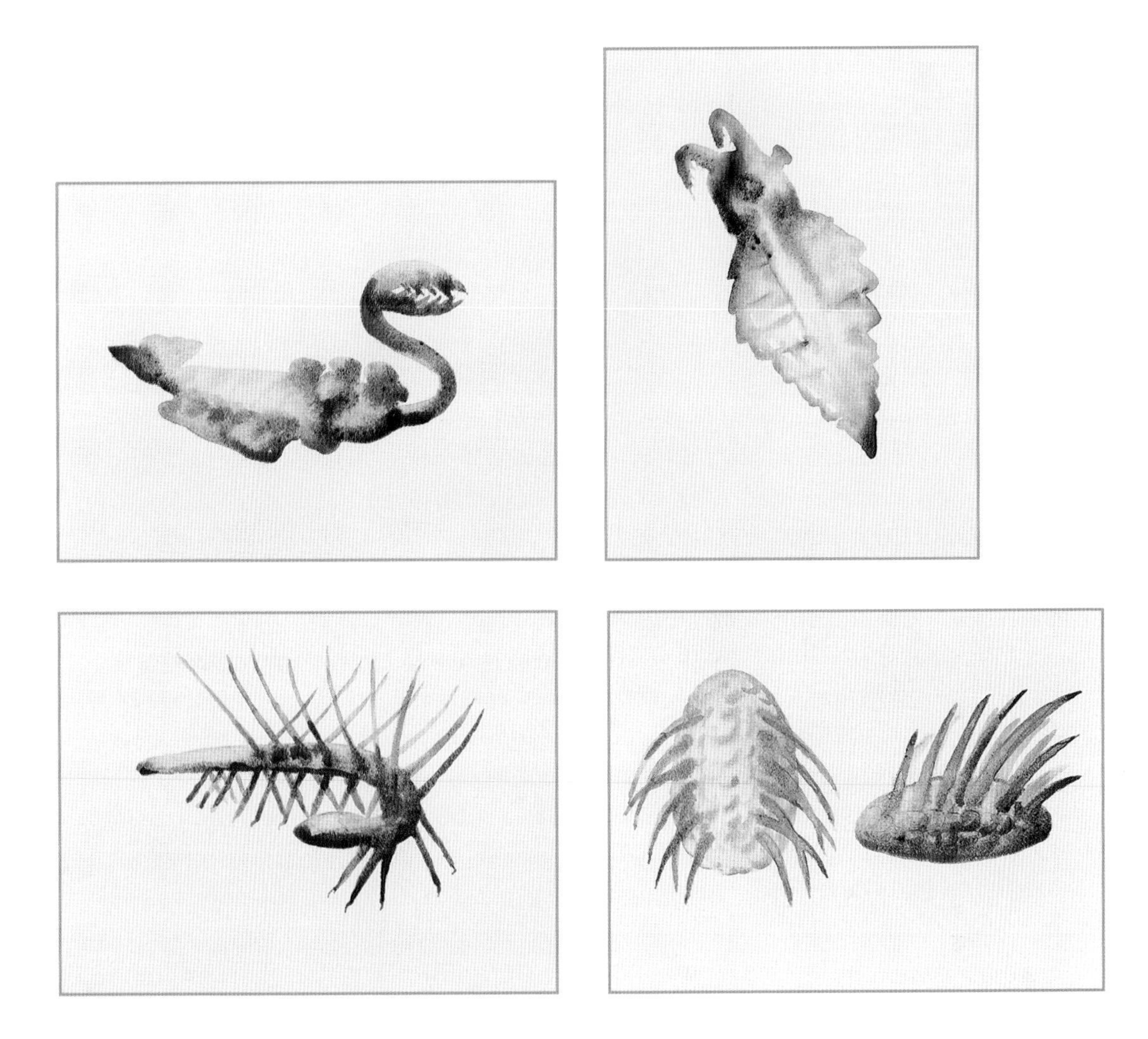

*The Cambrian Era saw an astonishing explosion of diverse new body plans.*

very few higher taxonomic categories. The evolutionary events of the Cambrian and the Early Triassic were dramatically different: although both produced a myriad of new species, the Cambrian event resulted in the formation of many new body plans; the Triassic event resulted only in the formation of new species that followed already well established body plans.

Two hypotheses have been proposed to explain this difference. The first supposes that evolutionary novelty comes only when ecological opportunities are truly large. During the Cambrian, for instance, there were many habitats and resources that were not occupied or exploited by marine invertebrate animals, and the great evolutionary burst of new body plans was a response to these opportunities. This situation was not duplicated after the Permo-Triassic mass extinction. Even though most species were exterminated, enough survived to fill most ecological niches. Under this scenario, there was sufficient survival of animals with various body forms to inhabit most of the various ecological niches (even if at low diversity or abundance) and in the process block evolutionary novelty.

The second possibility is that new phyla did not appear after the Permo-Triassic extinction because the *genomes* of the survivors had changed sufficiently since the early Cambrian to inhibit wholesale innovation. Under this scenario, evolutionary opportunities were available, but evolution was unable to create radically new designs from the available DNA. This is a sobering hypothesis, and one not easily discredited, for we have no way of comparing the DNA we find in living animals with the DNA from the long-extinct forms now preserved only as rock (movies such as *Jurassic Park* notwithstanding). It could be that genomes gradually become encumbered with ever more information—as they gather more and more genes—and in the process became less susceptible to critical mutations that could open the box of innovation.

One of the central—and currently controversial—aspects of the Cambrian Explosion concerns diversity and disparity. *Diversity* is usually understood as a measure of the number of species present. *Disparity* is a measure of the number of body plans, types, or design forms among those species. The controversy centers on the wondrous assemblage of fossils found at the Burgess Shale localities in western Canada, where not only early animals with hard parts, but also early forms without skeletons, are preserved as smears on the rocks.

The Burgess Shale has had an enormous impact on our understanding of the initial diversification of animal life. In large part, it is responsible for showing us that most or all of the various animal phyla (the major body plans) originated relatively quickly during the Cambrian. But the Burgess Shale may also be telling us that not only were the body plans found on Earth today around in the Cambrian, but so too were other body plans that are now extinct. One of the central messages of Stephen Jay Gould's book *Wonderful Life* is that the Cambrian was not only a time of great origination, but also a time of great extinction. Gould (and others as well) asserts that there were far more phyla present in the Cambrian than there are today. How many? Some paleontologists have speculated that there may have been as many as a hundred different phyla in the Cambrian, compared with the thirty-five still living today. In observing this pattern, says Gould, "we may acknowledge a central and surprising fact of life's history—marked decrease in disparity followed by an outstanding increase in diversity within the few surviving designs."

This view—so forcefully and beautifully described in Gould's *Wonderful Life*—is vigorously disputed in the 1997 book *The Crucible of Creation* by British paleontologist Simon Conway Morris, also about the Burgess Shale and the Cambrian Explosion. Conway Morris is, ironically enough, a central and sympathetic figure in Gould's book, which portrays him as one of the architects of our new understanding of the Cambrian Explosion. But he is not so sympathetic to Gould. He disputes Gould's assertion that disparity has been decreasing since the Cambrian, citing several cases suggesting just the opposite. Conway Morris also attacks another of Gould's ideas, the metaphor of "re-running the tape." Conway Morris argues that convergent evolution (in which distinct lineages evolve similarly in response to similar environmental conditions) can produce the same types of body plans from quite unrelated evolutionary lineages. He argues that even if the ancestor of the vertebrates had gone extinct during or soon after the Cambrian, it is likely that some other lineage would have then evolved a body plan with a backbone, since this design is optimal for swimming in water.

Simon Conway Morris's point is that convergent evolution will dominate evolutionary processes. He even makes what might be the first academic reference to Dougal Dixon's book *After Man*, a semi-whimsical prediction of how animals might look in the far future at a time when humankind has mysteriously gone extinct. Conway Morris notes that the animals conjectured by Dixon all seem to resemble animals living on Earth today, even though they are portrayed as evolving from quite novel sources:

> In the book he [Dixon] supposes that of all the mammals only a handful of types, mostly rats and rabbits, survived to repopulate the globe. *After Man* is

an exercise rich in imagination in its depiction of the riot of species that quickly radiate to refill the vacant ecological niches left after a time of devastation. All the animals, of course, are hypothetical. Certainly they look strange, sometimes almost alien. When we look more closely at their peculiarities, however, they turn out to be little more than skin deep. In this imaginary bestiary the basic types of mammal, those that trot across the grasslands, burrow in the soil, fly through the air, or swim in the oceans, all re-emerge.

It is thus the physics of environments that decides which shapes are adaptive and which are not, which shapes can allow flight or running ability or the ability to chase down and kill prey. And this assumption points to a central conclusion: no evolutionary novelty. Expectations of exciting and bizarre new life forms—the types seen in any science fiction B movie—are pipe dreams. The animals and plants arising in the future of evolution will in all probability look much like those of the present—except for being far less diverse.

## Expectations of Body Size

Large animals are much more charismatic than small ones. It is no coincidence that the majority of the animals listed as endangered and in need of help by the World Wildlife Fund are large mammals. Lately it has become fashionable to sneer at their popularity. Yet it is unfair to single out this hard-working group for criticism, for the hard fact remains that the large mammals—the last of the megamammals—are indeed endangered. Assuming that this group will show some of the highest extinction rates of the modern fauna, we might expect that future evolution will produce many new species of large-bodied animals.

Can we expect *any* new large animals? Recent scientific studies of the size distributions of mammals and their history of evolution suggest that large species might not appear after all. Mammalogists have long noted an interesting aspect of mammalian size distributions. There are over 4,700 species of mammals on Earth today, and their range in body size is impressive: the smallest (e.g., the tiniest shrews, such as the genus *Microsorex*) have an adult weight of about 2.5 grams, whereas the largest (*Balaenoptera musculus,* the blue whale) weighs about $1.6 \times 10^8$ grams—a difference of twenty orders of magnitude. And there are all size (weight) classes in between. Yet if the size distribution of mammals for each major continent on Earth is graphed, it is immediately apparent that each distribution is skewed to the right of the graph: there is a greater number of large mammal species than would be expected if the sizes were randomly distributed. However, this trend does not occur on small continents such as Australia, on large islands such as Madagascar, or on smaller islands. On smaller landmasses it turns out that the distribution of mammalian sizes is relatively symmetrical. Moreover, the smaller the landmass, the smaller the size of its largest mammals, and the larger the size of its smallest mammals. On small land areas the two tails of the distribution curve disappear. Finally, when small landmasses are completely isolated from other land areas, an extraordinary thing happens: large species evolve into dwarfs, and small species develop gigantism. But these are relative processes: an elephant evolving down to half size (the size of a horse) is still a large mammal (if a small elephant), whereas a mouse that doubles in size may be large for a mouse, but it is still a very small mammal. Can these observations be used to predict the future body sizes of newly evolving mammals (or any other type of animals, which presumably are affected in similar fashion)?

They can. Earlier we saw that global trade and travel are effectively recreating a supercontinent, bringing about the homogenization of the fauna that typified such large, single continents of the past. However, we have also seen that barbed wire, canals, roads, and freeways are subdividing the continents into smaller habitats. This trend is shaping the fauna according to the rules of island biogeography. Thus, we see the world being transformed, in an evolutionary sense, into an environment favoring low diversity, as well as the dwarfing of large species and the enlarging of small ones, with extinction occurring among the largest and smallest. The Age of Megamammals is well and truly over, with the last few wild megamammals now consigned to parks and zoos. As long as humanity survives at large population numbers, it will not return.

What might this new world look like? Let us invoke H. G. Wells's vehicle again for a fanciful, if dyspeptic, flight:

The Time Machine came to a stop. Ten million years had passed in the blink of an eye. The Time Voyager stepped from his machine and surveyed his surroundings. He was on the edge of a giant flat plain. Small fires dotted the broad expanse, sending thin blue columns of smoke into the cloudy and humid sky. The sun was setting, looking no different from the sun of his own time. Not for the first time, he wondered whether the machine had somehow malfunctioned.

Walking away from the Time Machine, he took better stock of his surroundings. He seemed to be in a gigantic garbage dump of some sort. Untold numbers of flies filled the air, their buzzing a constant hum of background Muzak. Roads crisscrossed the plain at regular intervals, but no vehicles could be seen. He looked more carefully at the refuse-strewn plain he was striding across, and was startled to see chaotic movement among the material swirling around him in the hot wind. At first he thought that thousands of huge insects were moving in the litter. But on closer inspection he discovered that while there were indeed numerous cockroach-sized insects afoot, many of the scurrying, tiny forms were mammals, a few as large as cats but most rat-, mouse-, or even shrew-sized.

He sat down on his haunches, motionless, and watched as more and more curious small mammals began to emerge. Clearly, many of the species he could now see were unfamiliar to him. Although their bodies looked like those of the rodents of his time, their heads were distinctly different. It was clear that many distinct species were present, some with long tapered heads, others with thin ribbonlike tongues, others with blunt heads and large knoblike teeth, still others with huge batlike eyes. Some had fur in a variety of camouflage patterns, while others were hairless. Some were heavily armored with armadillo-like scales. Some had front legs exquisitely adapted for digging; others had long needlelike claws extending from their toes. The small forms wormed among the garbage, some using their impossibly long tongues to probe into the piled refuse, while others broke open some of the many scattered bones to root out the marrow. While he watched, one of the small mammals was suddenly lifted into the air by a flicking rope of some sort, and then he saw the body being carried into a large, waiting mouth. A huge snake lay coiled not far away. Its tongue was like that of a frog, capable of flicking outward and grasping its prey. He saw another large snake moving on short legs like those of a centipede, and yet another moving its head in and out of the piled garbage, looking for small prey housed within.

Watching this menagerie of the small, he tried to compile a list of species new to him, losing count after tallying more than forty. It was not that everything was alien, for these creatures were surely descended from the mammals and reptiles of his own time, but they were just as clearly evolved, forming entirely new groups of species. And still more and more animals began to appear in the gathering twilight. The biggest animal that he saw was the size of a pig, and seemed indeed to be some sort of bizarre swine. But this "pig," if that is what it was, seemed well adapted for pushing through the piled garbage in search of food. It had a small trunk instead of a nose, which allowed it to root efficiently through the piles of rotting offal. Numerous small ratlike creatures hung from its sides, like remoras clinging to a shark. He thought at first that they might be babies, but they were clearly parasites of some sort, looking like hairy lampreys with greedy sucking mouths. Or perhaps vampires would be a better description. The rats, it seemed, had evolved.

He shuddered in revulsion at this bestiary. All seemed exquisitely adapted to these piles of garbage; in fact, all seemed adapted exclusively for life in this setting. In the distance he saw a copse of trees, and decided to leave the gigantic dump for a more "natural" setting, not realizing how natural garbage dumps were in this world. He began striding through the garbage, heading for the distant patch of green. Suddenly, shadows below and a cacophonous, raucous cawing from above announced a flight of birds over his head. They were crows, but bigger than those from his own time, and with brilliant plumage. He ignored them, but was jolted from behind with a sharp, piercing pain. Swearing, he put his hand to the back of his head, and found it covered with blood. He looked up to see another of the large crows diving at his head. He ducked just in time, seeing a large, eagle-like beak and great talons with a long, knifelike barb extending from one of the large feet. He began running back toward the center of the garbage, seeking shelter of some sort, but the crows, more than a dozen strong, attacked viciously. They let him run in terror back toward the trees, and as he got closer, he saw why: more than a hundred sat perched in the first row of trees, watching as their compatriots herded this particularly stupid human toward the waiting, hungry flock. The lions of the world now had wings.

*The most ubiquitous mammalian resource of the future: the human body.*

EIGHT

# THE FUTURE EVOLUTION OF HUMANS

> Some of our principal concerns about our bodies may be quite remote from, even at odds with, inherited human tactics for increasing our reproductive success or our chance of survival.
>
> —PHILIP KITCHER, *The Lives to Come*

What about the future evolution of our species? How will the future world and its new environments affect our own evolutionary outcome—or will we be affected at all? Will we become larger or smaller, or gain or lose intelligence, be it intellectual or emotional? Might we become more, or less, tolerant of oncoming environmental problems, such as a dearth of fresh water, an abundance of ultraviolet radiation, and a rise in global temperature? Will we produce a new species, or are we now evolutionarily sterile? Might the future evolution of humanity lie not within our genes, but in the augmentation of our brains through neural connections with inorganic machines? Are we but the builders of the next dominant intelligence on Earth—the machines?

Fossils tell us that evolutionary change is not a continuous thing; rather, it occurs in fits and starts, and it is certainly not "progressive" or directional. Organisms get smaller as well as larger, simpler as well as more complex. And while most lineages evolve through time in some manner, the most dramatic evolutionary changes often take place when a new species first appears. If this is the case, then future *morphological* evolution in *Homo sapiens* may be minimal. On the other hand, we may show radical change in our behavior and perhaps our physiology. Perhaps—and this is the biggest perhaps—a new species of human will evolve in the not so distant (or distant) future. But such an evolutionary change would almost surely require some sort of geographic isolation of a population of humans, and as long as humans are restricted to the surface of the Earth, such an event seems unlikely.

Since the time of Darwin, it has been accepted that the forces that produce new species are usually brought to bear when small populations of an already existing species get separated from the larger population and can no longer interbreed with it. Gene flow, the interchange of genetic material that maintains the integrity and identity of any species, thus gets cut off. Of course, genetic isolation generally means geographic separation, which means new environmental conditions, different from those experienced before. When you add this into the mix, you have a recipe for making a new species—given enough time, and continued isolation.

New species have formed many times in the course of human evolution. Although there are many gaps in the record (and disagreements among the specialists, with much work left to do), we can sketch a rough outline of human evolution. The human family, called the Hominidae, seems to begin about 4 million years ago with the appearance of a small proto-human called *Australopithecus afarensis.* Since then, our family has had as many as nine species, although there is ongoing debate about this number. About 3 million years ago two new species, *A. africanus* and *A. aethiopithecus,* appeared, while another, *A. boisei,* first appeared about 2.5 million years ago. (These three species are sometimes identified as *Paranthropus* instead of *Australopithecus.*) But the most important descendant of *A. afarensis* is the first member of our genus, *Homo,* a species named *Homo habilis* ("handy man") for its use of tools, an ability that is about 2.5 million years old. This creature gave rise to *Homo erectus* about 1.5 million years ago, and *H. erectus* gave rise to our species, *Homo sapiens,* either directly about 200,000 to 100,000 years ago, or through an evolutionary intermediate known as *Homo heidelburgensis.* Our species has been further subdivided into a number of separate varieties, one of which is the so-called Neanderthal. (Some researchers consider the Neanderthals to be a separate species, *Homo neanderthalis,* but this is still highly disputed.)

Each formation of a new human species occurred when a small group of hominids somehow became separated from a larger population for many generations. Then, following rapid morphological transformations, *Homo sapiens,* once evolved, showed *little or no further evolutionary change.*

Although major structural changes in *H. sapiens* may now be over, many smaller evolutionary changes will undoubtedly take place. Prominent among these might be a homogenization of the current human races. The same forces resulting in the homogenization of the Earth's biota are at work on us: our former geographic isolation has been broached by the ease of transportation and the dismantling of social barriers that once kept the *very* minor genetic differences of the various human racial groups intact. The most obvious change that may come about would occur in skin color. Because rapid transportation and global communication have destroyed most barriers to human movement and even isolationist human culture, we move about more. As we do so, we tend to interbreed, and thus the barriers that once selected for various types of skin pigment are no longer present. Skin pigment is one of the most heritable of human genetic features, and it may be that humanity is heading for a universal brown-skinned future, as the darkest of the black-skinned races get lighter and the melanin-free skins become darker. The humanity of ten thousand years from now might be but a single shade of color, a pleasing chocolate brown.

In stature, each race of our species seems to be getting larger, yet this is surely not a genetic feature: with improved nutrition we are simply maximizing the height potentials carried by our genes.

But in many ways, natural selection as we know it may not operate on our species at all. It is being thwarted on many fronts by our technology, our medicines, and our rapidly changing behavior and moral values. Babies no longer die in large numbers in most parts of the globe, and babies with the gravest types of genetic damage, which were once certainly fatal in pre-reproductive stages, are now kept alive. Predators, too, no longer affect the rules of survival. Tools, clothes, technology, medicine: all have increased our fitness for survival, but at the same time have thwarted the very mechanisms that brought about our creation through natural selection.

As an example of new human speciation, let's look for a moment at what it would take to create a new species with a much larger brain—say, a brain size of about 2,000 cubic centimeters, compared with the average value of about 1,100 to

1,500 cubic centimeters in *Homo sapiens.* What conditions of natural selection *on Earth today* would engender such a change, and would such a new creature even belong to our species?

## Intelligence and the Bell Curve

If *Homo sapiens sapiens* (the modern form of our species) were grouped together with an australopithecine, a *Homo habilis,* a *Homo erectus,* and an archaic *Homo sapiens,* what would the significant intellectual differences be? Would the other species use language, sing songs and create music, dream of flying, or even draw pictures? Before tackling these questions, we must first ask ourselves, what is intelligence?

There are several definitions of intelligence: what you use when you don't know what to do; guessing well about what fits together; finding an appropriate level of organization; finding an appropriate pattern from the available information. Although these statements certainly typify aspects of brain function that we recognize as intelligence, such definitions remain highly unsatisfactory. It is clear that intelligence is composed of an enormous array of components. Some of us have great math skills; most do not. Franklin D. Roosevelt, the longest-serving American president and certainly one of our best leaders, was an indifferent student. (Oliver Wendell Holmes once remarked of him, "A second-class intellect, but a first-class temperament!") His English contemporary and Second World War stablemate, Winston Churchill, was so indifferent a student that he never completed college and was packed off to the military by his despairing parents. Yet both rose to lead great countries in crisis through their *political* skills—clearly reflecting keen intelligence. Surely an ability to master politics is as much a type of intelligence as the ability to solve partial differential equations—but it is just as surely a very different type of intelligence. As any practitioner of "intelligence testing" can readily assert, the commonly used tests, such as the long-reigning IQ tests, measure some small portion of a great system of brain organization and function loosely termed "intelligence."

In the end, the definition of intelligence is probably irrelevant. For in spite of periodic hand-wringing by those who argue that our species is doomed to an ever-decreasing average intelligence because less intelligent people are having more children, there is very little chance of IQ (or any other measure of average intelligence) changing any time soon. The reason for this is that intelligence, by almost any definition, is produced by hundreds or even thousands of individual genes, and is thus very difficult to change. It is estimated that the correlation between an individual parent's intelligence and that of his or her child is 0.2%. Since both parents contribute, this effect is multiplied: thus the correlation between the parents' intelligence and that of their offspring is 0.04%. What this means is that two parents with IQs of 140 will probably conceive a child with an IQ of 100—as will two parents with IQs of 80.

Nevertheless, we remain fascinated by the concept of quantified intelligence, its history in our species, and the possibility of its long-term heritability. Those interested in the evolution of our species have probed the minutest anatomical differences among our various fossil predecessors in an effort to determine how and when our lineage began to "get smart." Yet this information is maddening in its incompleteness, and applying a present-day understanding of learning to the study of early humans is impossible. It has long been clear, for example, that the brain of a newborn human and the same brain only two to three short years later are vastly different, and the by-products of those differences are the marvelous characters of humanity. The toddler can speak in sentences, reason, remember, and move about independently; the infant can do none of these things. During the developmental period, and for years afterward, neurons are connecting and changing their morphology in ways still largely unknown to science. And no information about such changes is available to the paleoanthropologist when it comes to early humans. The most we can know of the brain of *Homo erectus* is its size and a bit about its shape, gleaned from the interiors of fossilized skulls. The real bits of information behind intelligence—the morphology of human brain cells and the pattern of their connections—is the area where the great secrets lie.

Two persons who have met with some success in this area are Terry Deacon of Boston University and William Calvin at the University of Washington. Deacon is a neuroanatomist who has studied the various attributes making up human intelligence. He has concluded that the emergence of human intelligence came about not through some mysterious new neurological or morphological invention within the brains of the earliest modern human species, but through the development of already-present circuits and cells. In other words, our species used "off-the-shelf equipment" that was then wired in novel ways through the processes of evolution.

Calvin, a neurobiologist, has put forward similar arguments about how intelligence came about through evolution. Calvin sees intelligence as the evolution of structural thought processes, such as syntax, the nested embedding of ideas; agendas, the ability to make novel plans for the future; logical chains of argument; and the ability to play games with arbitrary rules. And, last of all, Calvin sees a beautiful leap in the evolution of human intelligence: at some point, humans, alone among the animal world, began to perform and eventually write music. Calvin has presented a most interesting hypothesis about how all of this came about: he thinks that intelligence may be a by-product of a brain that evolved to throw better. So many new "connections" and novel pathways were necessary to create the neuroanatomy required for throwing weapons at prey that unforeseen consequences resulted from this newly evolved brain. In particular, it became intelligent.

## Unnatural Selection

In his book *Children of the Ice Age,* paleontologist Steven Stanley has made the observation that the advent of medicine has disrupted natural selection as it acts on humans. Humans, according to Stanley, have created *un*natural selection, for our species now routinely heals or saves many individuals who would never survive in the "wild." Furthermore, not only do we save individuals with physical or mental defects, we also allow them to breed. Now, with the increasing power of genetic engineering, we are poised to take unnatural selection to new levels, not only for a host of nonhuman species, but for ourselves as well.

One of the most provocative assertions about how our species is evolving at the present time comes from Dr. David Comings, a physician and geneticist specializing in human genetic disorders. In 1996 Comings published *The Gene Bomb,* a book as controversial (if far more overlooked) as *The Bell Curve.* Comings spent two decades studying Tourette's syndrome and attention deficit hyperactivity disorder (ADHD) among children, and came to the startling conclusion that the incidence of such genetically inherited disorders is increasing in the human population faster than population growth alone should dictate. His conclusion is that our species is evolving ever greater numbers of behavioral disorders.

Comings first came to this realization when treating his patients diagnosed with Tourette's syndrome and ADHD: he noticed that the frequencies of these disorders were high in the children of his patients. Instead of these behaviors being the result of an increasingly complex society, he surmised that society was selecting for the genes that caused these behaviors—a case of unnatural selection.

Further work now suggests than many of humanity's so-called "behavioral disorders"—such as ADHD, depression, addiction, and impulsive, compulsive, oppositional, and cognitive disorders—have a significant genetic component, *and unlike intelligence, may be coded on only a few genes.* If this hypothesis is correct (and no authoritative study has yet been able to falsify it), it means that the heritability (the rate at which such a trait is passed on to the next generation) of such disorders is very high. If selection acts to favor the transmittal of highly heritable traits, they very quickly increase in the overall gene pool of a species. Comings has summarized this stark view as follows:

> Many different studies have documented an increase in the frequency and a decrease in the age of onset of a wide range of behavioral disorders, including depression, suicide, alcohol and drug abuse, anxiety, ADHD, conduct disorder, autism, and learning disorders in the second half of the twentieth century. All of these disorders have a genetic component. The usual explanation of these trends has been that they are the result of an increasingly fast-paced and technologically complex society. I have suggested that the converse is true—that the increasingly complex society is selecting for the genes causing these behaviors.

Of course, these inherited behavioral traits are certainly affected by the individual's environment: many people carry the genes making them susceptible to addictive disease, but under many or even most circumstances do not succumb to alcoholism or drug addiction. Yet many others do. Comings took his findings even further, and postulates that many victims of ADHD reproduce at an earlier age than those without the syndrome, since very few of these sufferers attend college, and many women carriers become pregnant at an earlier age than women either attending college or entering skilled positions. The result is that women attending college, and then entering careers, ultimately have fewer children. These women are usually of higher intelligence than the population mean, and have lower frequencies of behavioral disorders. Women bearing children earlier—and having more children as a consequence—will be passing on their genes with more efficiency. While this difference will have little or no effect on intelligence, Comings believes that it may be highly significant in increasing levels of behavioral problems in the population.

Comings's theory is controversial for two reasons: the data and their interpretation. First, the increasing frequency of reported cases of depression, ADHD, and the like may simply be due to an increased awareness that these disorders can be treated, which encourages people to report them with greater frequency than they did in the past. Second, even if these disorders are indeed on the rise, they may have only a small genetic component, and may be due to any number of environmental causes, including increasing levels of environmental pollutants such as lead and other heavy metals, as well as organic macromolecules, in drinking water. Given these two issues, it is hard to say whether there is merit to Comings's assertions—but they provide an interesting example of the potential for further evolution in our species, involving genes that we usually do not think of as being capable of evolution.

## Human Behavior and Directed Evolution

We tend to think of evolution as something involving structural modification, yet it can and does affect things invisible to the morphologist—such as behavior. In fact, it may be that much of humanity's future evolution will involve new sets of behaviors, allowing us to deal with the changing set of environmental conditions facing our species: life in cities, life among crowds, life in a world where certain behaviors affect survival.

Because we have directed the evolution of so many animal and plant species, we might ask ourselves, why not direct our own? Why wait for natural selection to do the job when we can do it faster and in ways beneficial to ourselves? This is precisely the tack taken by many behavioral geneticists who are searching for ways to manipulate human genes. Behavioral genetics is a new branch of science that asks what in our genes makes us different from one another (vs. what makes us different from other species or what makes us human). Scientists working in this field are trying to track down the genetic components of behavior—not just of problems and disorders, such as those profiled in the previous section, but of everyday behaviors that may well be heritable traits: overall disposition, the predilection for addiction or criminality, many aspects of sexuality, aggressiveness, and competitiveness. These are traits that we know intuitively to be at least partially heritable.

The implications for the future of our species are incalculable. It seems unlikely that our society will not eventually accede to the idea that DNA samples should be given to genetic specialists. When this happens, elaborate screenings of an individual's genetic makeup will become commonplace, and specific genes for depression and other behavioral abnormalities will be detected. The second step will be the application of behavioral drugs using newly discovered chemical pathways. But the third step will be actual changes in people's genes. This can be done in two ways: somatically, by changing the genes in a relevant organ only; or by changing the entire genome—what is known as germ line therapy. Since germ line therapy involves changes in the genetic code of a person's eggs or sperm, it will not help the individual in question, but it will help his or her children.

The major obstacle to the genetic engineering of humans is a property known as pleiotropy: most genes perform more than one function, and many functions are coded on far more than one gene. All genes involved in behavior are probably pleiotropic. This is surely the case, for example, with the many genes involved in human intelligence (in fact, neuroanatomists and behavioral geneticists believe that the genes involved in IQ are probably involved in many basic brain functions as well). Therefore, far more will have to be known about the human genome before wholesale tinkering can begin, since very slight changes in gene frequency could lead to drastic changes in the species-level genome. As has often been quoted, a mere 1% difference in the genome is all that produces the vast gulf between chimpanzees and humans.

Why change genes at all, then? In all probability, the pressure will come from parents wanting to "improve" their children: to guarantee that their child will be a boy (or a girl), tall, beautiful, intelligent, musically gifted, sweet-natured, or wise, or to ensure that their children won't become addicts, thieves, mean-spirited, depressed, or hyperactive. The motives are there, and they are very strong. The Human Genome Project, now completed, had for much of its motivation (whatever it supporters argue) the desire to find "bad" genes. Once they are found, new Herculean efforts will be required to weed them out. Assuming that it does become practical to change the nature of our genes, how will that affect the future evolution of humanity? Probably a great deal, if the practice continues over millennia.

If natural selection is unlikely to produce a new human species—the event foreseen by H. G. Wells in his novel *The Time Machine*—the same end result could certainly be achieved by directed human effort. As easily as we breed new varieties of domesticated animals, we have it in our power to bring a new human race, variety, or species into this world. Whether we choose to follow such a path is for our descendants to decide.

Just as the push by parents to enhance their children genetically will be societally irresistible, the assault on human aging will be a force of unnatural selection in the future. Much recent research shows that aging is not so much a simple wearing down of body parts as it is a system of programmed decay, much of it genetically controlled. It is highly probable that the next century of genetic research will unlock numerous genes controlling many aspects of aging, and that these genes will be manipulated. An individual human lasting between one and two centuries is an obtainable goal. Whether or not it should be pursued, in light of human population growth, is another question.

Here is a scenario already posited by several scientists (and science fiction writers) that could potentially lead to a new human species, or at least a new variety. Some parents allow their unborn children to be genetically altered to enhance their intelligence, looks, and longevity. Let's assume that these children are as smart as they are long-lived—they have IQs of 150, and a maximum age of 150 as well. Unlike us, these new humans can breed for eighty years or more. Thus they have more children—and because they are both smart and live a long time, they accumulate wealth in ways different from us. Very quickly there will be pressure on these new humans to breed with others of their kind. Just as quickly, they will become behavioral outcasts. With some sort of presumably self-imposed geographic or social segregation, genetic drift might occur and, given enough time, might allow the differentiation of these forms into a new human species.

## Dyson among the Machines

Humans are no longer simply tool makers. Now we are machine makers as well, and not all of the machines we make can be considered tools. In ways perhaps even less predictable than our use of genetic manipulation, it may be our manipulation of machines—or they of us—that creates the most profound evolutionary change in our species. Not simple morphological change, or even behavioral change (although that might happen too)—but a change as consequential as the first enveloping of one bacterium by another to produce the symbiotic by-product now known as the eukaryotic cell—the key to animal life. Is the ultimate evolution of our species one of symbiosis with machines? Numerous writers have discussed the prospect, but in the late twentieth century, at least, perhaps none so evocatively as George Dyson, particularly in his book *Darwin among the Machines: The Evolution of Global Intelligence.*

The subtitle of Dyson's book sums up a possible trend in the future of our species. But according to Dyson, that global intelligence will not be a product of Darwinian evolution among the fusing populations of *Homo sapiens,* but will come about through an ongoing symbiosis with the machines we build: "Everything that human beings are doing to make it easier to operate computer networks is at the same time, but for different reasons, making it easier for computer networks to operate human beings."

In science fiction books and movies, this kind of symbiosis is portrayed as a machine and a man linked by cables—the contact of wire and neuron as a shared pathway of electrons. Would such a linkage enhance intelligence, if it were possible at all? Neuroanatomists claim that such linkages are only a matter of time and money, and that the first benefit of such a linkage will be enhanced memory—the ability for an individual to immediately access the knowledge of the collective. But is memory—and data—intelligence? Dyson points out that H. G. Wells pondered this subject in his lifetime, concluding that some sort of global intelligence was the only hope of improving the affairs of humanity. Wells prophesied that the whole human memory would soon be accessible to every individual. In Dyson's view, "Wells acknowledged memory not as an accessory to intelligence, but as the substance from which intelligence is formed."

All of us who routinely use computers have suffered from some lack of memory in our systems, be it RAM or space on a hard disk drive, and such nuisances invariably detract from some other task, break our concentration, require an unanticipated change in activity. It is easy to see how extra memory space—or extra memories—would change the world and the way we perceive it. But would it increase intelligence? Most thinkers who ponder this subject assure us that it would, though in ways that may not be perceptible to us, perhaps because an enlarged, networked intelligence would operate at time scales different from ours, and thus invisible to us. As Dyson notes, it might also operate in a fashion unlike that of Darwinian evolution:

> What leads organisms to evolve to higher types? Darwinian evolution, as Stephen J. Gould, among others, has pointed out, does not "progress" toward greater complexity, but Darwinian evolution, plus symbiogenesis, does. . . . Darwinian evolution, in one of those paradoxes with which life abounds, may be a victim of its own success, unable to keep up with non-Darwinian processes that it has spawned.

In an earlier chapter, we asked whether the "rules" of speciation have changed for humans. The answer is that they have not—humanity may have affected the nature of the playing field and the players, but we cannot change the rules. Yet in the *merging of man and machine* that conclusion may be overturned. The evolution of machines and machine intelligence *does* have directionality, toward ever greater complexity and intelligence. Machine intelligence does not go backward as much as it goes forward; there are no functional equivalents of a blind cave fish or the whale, a mammal that returned to the sea. In the computer world, direction is progress: better operating systems, more machine interconnections, more memory, easier use, more humans connected.

Dyson further argues that information comes in two types: structure and sequence. The first is the map of space, the second the map of time. Memory and recall are translations between these two types of bits. Thus it is Dyson's fervent belief that the future evolution of humanity lies in "technology, hailed as the means of bringing nature under the control of our intelligence, thus enabling nature to exercise intelligence over us. We have mapped, tamed, and dismembered the physical wilderness of our Earth. But, at the same time, we have created a digital wilderness whose evolution may embody a collective wisdom greater than our own."

As breathtaking as Dyson's vision is, I differ slightly in predicting the type of machines we may merge with. One of the tired old saws in the science fiction pantheon is the notion of a silicon-based life form. There is a simple but powerful retort to that possibility. The variety of "organic," or carbon-based, compounds found in life processes can be readily seen by going to any chemical supply store and checking out the store's catalogs. They are book-sized. Silicon-based compounds, on the other hand, fill only a good-sized comic book. Silicon, however useful it may be for the electronics and computer industries, is none too suitable for life. We may soon find that it is none too suitable for the computers of the future, either, or for the machines we may try to merge with.

Perhaps, after all, the progression of Earth's dominant animals will be Age of Bacteria, Age of Protists, Age of Invertebrates, Age of Fishes, Age of Amphibians, Age of Reptiles, Age of Mammals, Age of Humanity—and then a long Age of Artificial Intelligence. This certainly seems to be the view of many of the moguls and thinkers spawned in Silicon Valley. Of these many prophets, none seems so bullish concerning the coming replacement of humanity by thinking machines as Ray Kurzweil. His vision is starkly writ in *The Age of Spiritual Machines.* Kurzweil believes that the invention and routine construction of computers that have the computing power of the human mind will occur early in the twenty-first century, and that machines that outstrip the capabilities of the human brain in some attributes of processing and logic will inexorably appear soon after. In his view, the merging of human and machine (or at least artificially constructed) brains will be inevitable. But will it ever be heritable?

## Growing Buildings and Changing Trophic Levels

Science fiction is so pervasive and voluminous in its output that there are few ideas not already in use in some futuristic plot device. Thus, the two ideas I throw out here are surely well known to its aficionados. However, they seem to me to be two potentially realistic strategies that could alleviate both our population problems and the stresses our metals-based industries place on the rest of the world.

Many of our industrial problems and pollutants come from the processing of metals. The industrial smelting and forging of tools and technology made from iron, aluminum, nickel, tin, copper, and the many minor metals and their blended components requires massive supplies of energy and water and produces volumes of pollution. Although it would require significant genetic manipulation to create organic structures and tools, humanity might be better off by "growing" as much technology as possible. On a very small level, researchers have already experimented with this idea: a square tomato has been developed to suit packaging needs. More imaginatively, a living house made of growing wood and other organic structures might be a way to realize sustainability in a technical society.

Yet an even more dramatic breakthrough could be realized by manipulating not only our machines and technology, but also ourselves. Humans require massive amounts of food. We, like all animals, are *heterotrophs*—we must ingest other once living matter in order to live. The world could support far more people if we could somehow radically re-engineer our food and nutrient needs so as to become *autotrophs*—organisms at a lower trophic level. Plants and many types of chemoautotrophic bacteria use sunlight or chemical energy to power their metabolism. If biotechnology could help get humanity off its hamburger diet (or even its wheat diet) and merge with the plant world, a great deal of stress on the planet could be alleviated. Solar-powered calculators do remarkably well; perhaps the only hope for an Earth even remotely resembling its past self in terms of habitats and diversity is solar-powered humans.

## New Human Species?

Our lineage has produced new species in the past. What about the future?

Speciation requires an isolating mechanism of some sort. The most common is geographic isolation, whereby a small population gets cut off from the larger gene pool, then transforms its own set of genes sufficiently that it can no longer successfully reproduce with the parent population. Most species have done this through geographic isolation, yet the very population size and efficiency of transport of humanity make this possibility remote—at least on Earth. If, however, human colonies are set up on distant worlds, and then cut off from common gene flow, new human species could indeed arise.

Perhaps humans will lose (or voluntarily discard) the technology that allows the global interchange of our species from continent to continent. If separation lasts long enough, and if conditions on the separated continents are sufficiently different, it is conceivable that a new human species could arise due to geographic isolation.

## Scenarios

Let us conclude with some alternative scenarios of humanity in the future. Once we leave aside our guilty assumption that our species will soon go extinct somehow, we are left with what Rod Taylor (in the film version of H. G. Wells's *The Time Machine*) described as "all the time in the world." With hundreds of thousands to millions of years yet to play with, what might our species evolve into? Here are four scenarios:

1. *Stasis:* In this scenario we largely stay as we are now: isolated individuals. Minor tweaks may occur, mainly through the merging of the various races.
2. *Speciation:* Through some type of isolating mechanism, a new human species evolves, either on this planet or on another world following space travel and colonization.
3. *Symbiosis with machines:* The evolution of a collective global intelligence comes about through the integration of machines and human brains.
4. *Eusociality:* Our fascination with ants is that we see our cities and ourselves mirrored in them. The animal world is filled with colonial organisms. Hydrozoans and bryozoans have morphologically distinct polyps that serve for food acquisition, defense, reproduction, and colony stabilization. Each

> polyp is connected to every other polyp. The functional equivalent of this system among insects is the behavior of species like ants, known as eusociality. Ants (and other eusocial insects) have evolved behaviors and morphology befitting a highly complex system in which the colony itself serves as the functional individual, and the various actual ants of the colony serve as the various organs of that "superorganism." Will the future evolution of our species be toward the ant model? In one of the most original of all science fiction novels, Larry Niven and Jerry Pournelle's *The Mote in God's Eye,* an intelligent race genetically manipulates itself to evolve different types of "workers," including morphologically discrete farm workers, engineers, politicians, soldiers, "masters," and even "food."

Evolution takes time. Humanity probably has that—as much as a billion years of it—as I will show in the next two chapters.

*It is unlikely that even geological boundaries will apply to humans in the future.*

NINE

# SCENARIOS OF HUMAN EXTINCTION

## Will There be an "After Man"?

Ultimately the Earth could no longer supply the raw material needed for man's agriculture, industry or medicine, and as shortage of supply caused the complex and interlocking social and technological edifices crumbled. Man, no longer able to adapt, rushed uncontrollably to his inevitable extinction.

—DOUGAL DIXON, *After Man*

On May 23, 1995, academician Boris Pevnitsky, Deputy Director of the Russian Federal Nuclear Center, strode up to a small podium in a cramped, steamy room in Livermore, California. His audience consisted of an attentive, cosmopolitan group of middle-aged men: some in sleek Nordstrom's suits, some in the more dingy variety available at GUM in Moscow, others in United States Air Force uniforms. Dr. Pevnitsky's message was simple: he proposed that the United States and Russia jointly build the largest hydrogen bomb ever conceived, a bomb so great in energy yield that its explosion anywhere on Earth would blow away a significant portion of the atmosphere. But that was Dr. Pevnitsky's point. He didn't intend to explode his "classical super" anywhere near the Earth. He intended to use it to destroy an asteroid in space. In the back of the room, a small, shriveled man with graying but still fierce eyebrows looked on, surely with some satisfaction. Dr. Edward Teller, inventor of the H-bomb, must have been pleased to hear that his great dream—the "peaceful" use of thermonuclear weapons—was at last being discussed by an international assemblage of American, Russian, and Chinese nuclear specialists, astronomers, and geologists, 150 strong, in an open forum. All had gathered in the slightly run-down conference center of the Lawrence Livermore Laboratory to discuss one topic: the defense of the Earth from comet or asteroid bombardment.

The Lawrence Livermore meeting was the second such meeting held in a one-month span in 1995. The first, held at the United Nations in New York in late April of that year, had roughly the same theme, but was far more theoretical, dealing with the rate of asteroidal collisions, rather than how to defend against them. The megatonnage necessary to deflect or destroy an asteroid was not overtly discussed, but the message was the same: our planet, and with it every species, every individual, every great and tawdry work of humanity, is endangered by celestial happenstance. The magnitude of this risk remains a topic of utmost importance to our species, for many scientists believe that the greatest threat facing us is not some Hot Zone virus from Africa, but the billion or more comets and asteroids that have Earth-crossing or potentially Earth-crossing orbits. Should Halley's comet, or some other messenger from the heavens of equal size, hit the Earth, it would bring about the complete destruction of all life on this planet.

Asteroid impact is not the only threat facing our species, only the single most dangerous one. Other threats to the viability of the human species exist as well, including other astronomical disasters (a gamma ray burst, a nearby supernova) as well as Earth-borne causes such as thermonuclear war, biological warfare, disease, and sudden climate change. While any of these disasters would severely reduce human populations, probably none by itself, other than the astronomical causes, is capable of driving all of humanity to extinction. However, several of these effects acting in tandem could do the job.

## The Sky Is Falling! The Sky Is Falling!

Despite its scientifically ludicrous depictions, Hollywood has nevertheless served to educate people about the dangers of Earth-crossing asteroids and comets. The overall message of the late-twentieth-century movies *Armageddon* and *Deep Impact* was grounded in reality. Asteroid impact can be viewed as the most dangerous single threat to our species' existence.

The rate of collision of celestial objects with the Earth has been well established, and the destruction wreaked by such impacts is also well understood. Such events have occurred countless times throughout the history of our planet—and they will inevitably recur in the future. For example, it was a great planetary collision nearly 4 billion years ago that created the Earth-moon system, and in so doing may have made our planet unique as a womb for the gestation and diversification of life. A second great collision some 65 million years ago slayed the planet's dragons and set the stage for mammalian, and ultimately human, evolution. But of greatest relevance to our species are the future collisions, great and small, that will inevitably occur. For alone among the countless species that have populated this Earth, we have it in our power to defend the planet from these strikes.

Just how likely is it that a comet or asteroid collision might destroy our civilization? Is it more dangerous to arm ourselves with nuclear weapons larger and more destructive than any yet developed, ostensibly for planetary defense, than it is simply to pray?

While it is impossible to assign a precise number to the chance of an asteroid impact, we do know that significant hits have occurred as recently as 100 years ago. On June 30, 1908, a relatively small meteor exploded in the lower atmosphere over remote Siberia, releasing the force of a hundred Hiroshima-sized atomic bombs. The blast flattened miles of forest, and 50 miles away, reindeer herders and their stock were blown into the air. Had that same explosion happened over a heavily populated area, it would have produced one of the greatest episodes of human carnage in recorded history. The small fragment that produced this explosion was only about 50 meters across. Just two years ago, an asteroid a hundred times as large barely missed the Earth, and was seen only *after* it went by. Today, several hundred Earth-crossing asteroids of various sizes have been detected, and that seems to be only a small sampling of the thousands that orbit in the vicinity of Earth, not to mention the estimated billion comets hovering far out in space. If an object larger than a mile in diameter were to strike the Earth, not only would our civilization be threatened, but so too would our species. And as we saw with the impact of Comet Shoemaker-Levy 9 on Jupiter, the time between detection and impact for even a giant comet—a species-killing comet, had it struck Earth—can be less than a year.

The reality is that this planet will continue to be hit by debris from space. Just as with earthquakes, the question is not if, but when—and how big. Very large impacts—such as those that might bring about mass extinctions—seem to occur at intervals of tens to hundreds of millions of years. But the frequencies and ages of craters found on the surface of the Earth demonstrate that asteroids of up to a kilometer in diameter seem to hit the Earth at million-year frequencies. Such collisions could be expected to disrupt human agriculture on a global scale for many years following the impact, and would certainly lead to a great slaughter of humanity. The complete extinction of humanity might occur through the impact of a comet or asteroid greater than about 15 kilometers in diameter—half again as large as the one that ended the Mesozoic era (and killed off the dinosaurs in the process) some 65 million years ago.

Although the impact of a very large asteroid or comet could completely eliminate the human race, a more likely scenario is that some significant proportion of our population would be removed by the direct effects of such an event, and the rest of the job would be done by the aftereffects. Just how easily that could occur was shown by astronomer John Lewis of the University of Arizona in his 1999 book, *Comet and Asteroid Impact Hazards on a Populated Earth.* Lewis not only wrote about the dangers of such events, but included with his published book a software program that allows the reader to simulate such impacts.

Lewis's simulation program uses a statistical analysis to calculate the human deaths resulting from an impact. Michael Paine, a scientist at the Jet Propulsion Laboratory at Pasadena, ran the program to simulate the effects of likely asteroid impacts on human populations over the next million years. His analysis led to sobering results. Assuming a constant population of 5 billion people on the Earth at any time, the total death toll was 7.5 billion people over a million-year time span—or 7,500 fatalities per year. But the impacts were not evenly distributed. Paine's simulation yielded ten occurrences in which the impacting body was from one thousand yards to a mile in diameter, five in which it was a mile to 1.3 miles in diameter, and one in which the impacting body was greater than 1.3 miles in diameter. The latter event resulted in 2.5 billion deaths when a 1.3-mile comet hit the

American Midwest, releasing the energy of 60,000 H-bombs. Seven million people were killed instantly, and the rest died of starvation as sunlight was blocked and crops failed.

If our species survives for a long period of time—as I believe it will—then we will become used to occasional devastating global events that—at a minimum—knock humanity back into the Stone Age for long periods of time. One such event in conjunction with another human-killing factor could indeed cause our extinction. Of course, the impact of an even larger asteroid or comet could do the job all by itself.

## War and Thermonuclear War

Although it is doubtful that thermonuclear war alone could wipe out humanity, one scenario for human-induced extinction is a massive thermonuclear exchange, perhaps aided and abetted by chemical and biological warfare at the same time. While the causes of such warfare could come from many sources, the twenty-first century and beyond will probably see a variety of smaller wars fought over food, land, and water. Unto themselves these wars certainly pose no threat to the entire species—unless, of course, they escalate into a full-blown nuclear-chemical-biological exchange.

Human beings have proved to be a cantankerous lot, and unfortunately the efficiency of anti-human weaponry has markedly increased since our stone-throwing days. Prior to the twentieth century, war was an exercise between combatant armies. In that century, however, it spread from armies to noncombatant populations. William Eckhardt, in his "War Related Deaths Since 300 B.C.," has estimated that the number of war deaths per 1,000 people in the total world population rose from 9.7 in 1700–1799 to 16 in 1800–1899 and to 44.4 during the twentieth century. This change was promoted by the twentieth century's propensity for attacking an enemy nation's economy, infrastructure, and civilian population. Total war deaths for the twentieth century totaled nearly 110 million humans, at least half of whom were civilians. Yet alarming as this rising death toll is, our extinction would require far more than 45 deaths per thousand.

The reasons for future conflicts are easy to discern. First, the need to supply food to a growing human population has caused a great increase in the need for arable land and for water to irrigate crops. New high-yield strains of rice and other grains, for example, require more water than less productive crops. Often, aquifers that have taken centuries or millennia to fill take only years or decades to discharge. One solution to water problems has been to build dams, but damming rivers that travel through more than a single country is a sure means of creating conflict. From 2000 to 2020 the worldwide demand for water is expected to double due to the need to irrigate new agricultural lands in dry areas, as well as the need to provide water for the growing human population and for industrial uses. There will also be renewed demand in dry countries such as Jordan, Israel, and Syria, where aquifers have been nearly emptied. Not surprisingly, in the late twentieth century, the World Bank prophesied that most twenty-first-century wars would be fought over water supplies.

Whatever its cause, war has been a plague on humanity as long as there has been human memory and legend. Prior to the latter half of the twentieth century, however, none of the myriad means of human warfare posed a threat to the species. But the unleashing of the atom and of fusion reactions has changed all that. Today, humanity holds the seeds of its own destruction in its hands.

Following the explosion of the first atomic bomb in 1945, and the first thermonuclear (or hydrogen) bomb in the early 1950s, the stockpile of such weapons has grown alarmingly. The National Research Council estimated that the five major nuclear powers (the United States, Russia, Britain, France, and China) possessed nearly 70,000 nuclear weapons in the mid-1980s. This number had diminished by the latter part of the 1990s to a total of about 35,000 weapons, with about 23,000 of these deployed at about 90 sites in Russia. While a smattering of other weapons exist, notably in Israel, India, and Pakistan, it is the two superpowers that maintain the majority of the arsenal's numbers. The total is expected to decline further, to about 6,500 active weapons for the United States and Russia combined by 2003—roughly the same number present in 1969.

While the reduction of weapons systems has somewhat diminished the overall danger, the truth of the matter is that there are far more nuclear weapons still on line than would be necessary to exterminate humanity from the Earth—one instance in which the word "overkill" is not hyperbole. Scenarios of nuclear exchanges include vast devastation and radiation poisoning, with long-term aftereffects due to the lengthy half-lives of the radioactive materials released into the environment. In an article published in 1982, Paul Crutzen and John Birks pointed out a further peril, later dubbed "nuclear winter." They suggested that multiple nuclear explosions could blacken the atmosphere with enough soot and dust to reduce sunlight by 99% for a period of 3 to 12 months, depending on the number, yield, and type of target of the exploding warheads. Such a cloud cover could reduce average global temperatures to well below freezing in the interiors of North America and Asia. Such temperature changes, while not necessarily dooming humanity, would certainly reduce agricultural crop yield to a trickle.

The concept of nuclear winter gained credence from studies of the terminal Cretaceous asteroid collision and from additional work by Carl Sagan and his colleagues. They showed that a full-scale nuclear exchange could indeed bring on a "nuclear winter," which in turn could quite easily lead to the extinction of the human species.

## Catastrophic Climate Change

Never in the history of life on Earth has there been an organism better adapted for climate change than the human species. The most serious environmental threat to most species is temperature change, but we can deal with that quite easily: if too hot, remove clothes, or install an air conditioner, or move to a cooler climate if necessary. Dealing with cold is even easier: put on warmer clothes, build a fire, install insulation: in short, let technology deal with the problem. No other animal has the ability to control its body temperature so quickly and easily. Thus temperature change is usually not a threat to our species' survival.

Or is it? One of the great scientific discoveries of the latter part of the twentieth century has been a new understanding of the rate of climate change in the past. It had long been assumed that major changes in the Earth's climate were drawn-out events spanning great intervals of geologic time. Evidence to the contrary began to emerge when deep cores of ice drilled from the ancient ice sheets covering parts of Greenland and Antarctica were analyzed for their isotopic content. To everyone's surprise, these long records of global climate revealed intervals of extremely rapid temperature change. The newly discovered data painted a much more dramatic picture of climate change: they showed that major changes could overtake the Earth in a decade or less. And the changes would not be limited to the high north or south—their effects would be global. Such rapid changes, if superimposed on the large human population and its present agricultural needs, would be a recipe for chaos and at least partial extinction of our species.

Somewhat paradoxically, it may be that global warming will trigger a rapid change involving a sudden cooling of the Earth. In a thoughtful article published in the *Atlantic Monthly,* William Calvin of the University of Washington outlined a scenario that, if it did not exactly drive our species into extinction, could certainly set up the social chaos that would lead to global war and the loss of significant portions of the human population in mere decades. Calvin called this scenario the "great climate flip-flop."

Calvin argues that catastrophe could come with a sudden cooling of Europe. At the present, Europe is anomalously warm for its latitude. Whereas most of the populous parts of North America lie at latitudes between about 30° and 45°N, most of the population of Europe is about ten degrees farther north: London and Paris are at nearly 50°N, Berlin at 52°N, Copenhagen and Moscow at 56°N, and the cities of Scandinavia at 60°N. Yet in spite of its more northerly location, the European subcontinent is extremely productive. Its agricultural industry supports twice the human population of North America on a much smaller landmass. Much of its agricultural success comes from a climate warmer than its latitude might otherwise dictate. Europe's warmth comes from the Gulf Stream, a tropical water current flowing up the Eastern Seaboard of North America and then vaulting across the Atlantic to push masses of warm water against the European landmass. It keeps northern Europe about 10° to 20° warmer than it otherwise would be.

The current bringing heat to Europe is not a single waterway, but is composed of several segments. One branch of this current carries warm water to the vicinity of Iceland and Norway. Eventually this water cools, and when it does, it sinks deeper into the ocean. It then returns to more southerly latitudes, but does so as a cool deepwater current, rather than the warm surface current it begins as. As it moves south it also carries more salt with it, for salt water is heavier than fresh water and tends to sink because of its greater density. Warmer, fresher water thus travels on the surface and returns at depth as cooler, saltier water. Paradoxically, this system would be shut down if more fresh water were added to it on the sea surface. Thus the movement of salt in the current system is integral to maintaining the steady supply of warm water to the coast of northern Europe.

The scenario that could lead to a failure of the warm northern Atlantic current is global warming. If the glacial ice covering Greenland were to melt at a higher rate than it currently does, it would flood the sea surface in the regions concerned with fresh water. The normal circulation pattern would thus be disrupted, causing the northern branch of the current to begin its return prior to reaching Greenland. The warm temperatures that these currents bring would not reach the shores of Europe, and Europe would suddenly cool. Thus a strange paradox: global warming would ultimately cause a sudden cooling of Europe.

The failure of one single current would, at first glance, not seem to be the stuff of sudden global climate change. Yet the world's oceans are but a single body of water, and heat flow is global. The breakdown of any one current system necessarily causes changes in others. If the North Atlantic current failed, the entire world would experience sudden climate change. Europe would go into a deep freeze, and it seems likely that its human population would revert to warfare over the territory necessary to support the suddenly starving millions. It may sound alarmist and overly dramatic to talk about "suddenly starving millions," but it is important to remember that Europe currently has 650 million people and is largely self-sufficient in its food production. Almost simultaneously with the global current change, that ability to be self-sustaining would disappear. Calvin describes this scenario as follows:

> Plummeting crop yields would cause some powerful countries to try to take over their neighbors or distant lands—if only because their armies, unpaid and lacking food, would go marauding, both at home and across the borders. The better organized countries would attempt to use their armies, before they fell apart entirely, to take over countries with significant remaining resources, driving out or starving their inhabitants if not using modern weapons to accomplish the same end: eliminating competitors for the remaining food. This would be a world-wide problem—and could lead to a Third World War.

Calvin makes the point that without its warming current, Europe would have a climate like that of present-day Canada, and if Europe had weather like Canada's, it could feed only one of twenty-three of its inhabitants.

What makes a sudden global cooling especially threatening is that it is not a "point source" disaster. The Earth is regularly stricken with calamities such as hurricanes, tornadoes, and catastrophic earthquakes. These disasters cause the loss of much human life, and they are usually followed by rescue and recovery efforts that are often global in scale. But such disasters are always of short duration and of limited geographic extent. Neither hurricanes nor earthquakes perturb any significant percentage of the Earth's surface for more than a few days. An abrupt cooling, on the other hand, could last decades or centuries. Calvin argues that even a meteor strike killing a majority of the human population in a short period of time would not be as catastrophic as a longer-term disaster that killed just as many—the killing effects of the meteor strike would soon be over, but global cooling would continue to stretch its deadly effects over decades at a minimum, and more likely over centuries.

## Disease

In the last years of the twentieth century great attention was centered on human communicable diseases, sparked by a spate of movies and best-selling books. What are the chances that a new disease could bring about the extinction of humanity? For instance, what if a 100% fatal disease such as HIV were spread as readily as the common cold? And what if such a disease were used as a weapon? Biological warfare, like nuclear warfare, does have the potential for a radical reduction of the human population if world war erupted. Most disturbing may be the stockpiling of diseases for which we no longer are vaccinated (e.g., smallpox), and the genetic engineering of new, virulent strains of disease, the subject of countless movie and book plots.

Two observations argue against the possibility of a species-ending epidemic. First, there is no evidence that any single disease has ever killed off any species. Second, humanity now has a technology that can combat disease with increasing efficiency each year. Nevertheless, disease remains a potent method of reducing human population numbers, and when combined with (or resulting from) other human killers in synergy, it could well be a potent force leading to the extinction of our species, especially if global warming causes tropical diseases to move into previously temperate regions.

## Death from the Robots

It is difficult to arrive at any scenario of human extinction (or any scenario of anything, for that matter) that has not already been featured in some Hollywood movie. So too with our next potential villain, artificially constructed machine intelligence. In the famous *Terminator* and *Terminator 2* movies (and to some extent in *The Matrix* as well), the near-future world is run by malevolent robots that are trying to exterminate the human species, or at least enslave it. Such a scenario seemed highly likely to Ted Kaczynski, the infamous Unabomber, who in his manifesto published by *The New York Times* and *The Washington Post,* wrote, "Let us postulate that the computer scientists succeed in developing intelligent machines that can do all things better than human beings can do them. . . . If the machines are permitted to make all their own decisions . . . the fate of the human race would be at the mercy of the machines."

This view is also expressed by roboticist Hans Moravec, in his book *Robot: Mere Machine to Transcendent Mind,* and by Bill Joy, the co-founder and chief scientist of Sun Microsystems, in his chilling 2000 article *Why the Future Doesn't Need Us.* As Moravec says, "In a completely free marketplace, superior robots would surely affect humans. . . . Unable to afford the necessities of life, biological humans would be squeezed out of existence." Moravec foresees the fusion of a human being and a robotic body or being to produce a superintelligent hybrid of some sort (as always, Sci-Fi has been there, done that, the Borg from *Star Trek* being only the most recent entry into this genre). To Moravec, such a being would succeed humanity—and cause our extinction sooner or later.

Why robotics? Part of the promise is a better way of life for the organic makers—us. To do away with the mind-killing labor that bedevils most of humanity would indeed be a social and intellectual breakthrough. But robotics holds an even greater promise—the extension of our individual intellects, if not our bodies. If we can download our consciousness into a machine (and by machine I mean organically produced as well as purely inorganic artificial intelligence), we will indeed be on the verge of some sort of immortality. But at what cost? As Bill Joy notes, "If we are downloaded into our technology, what are the chances that we will thereafter be ourselves or even human? It seems to me far more likely that a robotic existence would not be like a human one in any sense that we understand, and that robots would in no sense be our children, that on this path our humanity may well be lost."

## Gray Goo

Of all threats facing humanity, perhaps none is so dangerous—or so poorly understood—as that posed by nanotechnology. *Nano* means "small," and many technologists now see the future of technology as the manipulation and assembly of matter at molecular and even atomic size scales. Such molecular-level assemblers could utterly transform human society by creating extremely inexpensive products, medicines, and even energy through the construction of virtually free solar panels. Because most of the new products would be created from organic rather than metallic and other mineral material, there would be far less pollution and fewer other environmental consequences of manufacturing.

The future in such a world might be utopian. On the other hand, it might mean the extinction of humankind. It is this vision that is most starkly illuminated in Bill Joy's cautionary 2000 essay. Joy views nanotechnology as the most dangerous of the new trio of technologies, dubbed GNR, for *G*enetics, *N*anotechnology, and *R*obotics. As he notes, "Molecular electronics—the new subfield of nanotechnology where individual molecules are circuit elements—should mature quickly and become enormously lucrative within this decade, causing a large incremental investment in all nanotechnologies. Unfortunately, as with nuclear technology, it is far easier to create destructive uses for nanotechnology than constructive ones. Nanotechnology has clear military and terrorist uses."

Joy sees the military applications of nanotechnology as potentially dangerous to the existence of our species. Moreover, just as the nonmilitary use of atomic energy holds undeniable threats to human life from nuclear power plant accidents, so too does the potential exist for industrial accidents in commercial nanotechnology applications. Yet while one cannot imagine any scenario in which the release of radioactivity from an industrial application of nuclear power would threaten the entire human species, a "runaway" nanotechnology could. Such a case is described in Eric Drexler's book *Engines of Creation:*

> Tough omnivorous bacteria could out-compete bacteria; they could spread like blowing pollen, replicate swiftly, and reduce the biosphere to dust in a number of days. Dangerous replicators could easily be too tough, small, and rapidly spreading to stop—at least if we make no preparation. We have trouble enough controlling fruit flies.

Among the cognoscenti of nanotechnology, this threat has become known as the "gray goo problem." Though masses of uncontrolled replicators need not be gray or gooey, the term "gray goo" emphasizes that replicators able to obliterate life might be less inspiring than a single species of crabgrass. They might be superior in an evolutionary sense, but this need not make them valuable. The gray goo threat makes one thing perfectly clear: we cannot afford certain kinds of accidents with replicating assemblers.

## No End?

Although the litany of dangers facing our species seems daunting, none is an unambiguous death sentence. Each can be dealt with if our species shows foresight. These dangers must confront any race that climbs the evolutionary ladder to intelligence. As Carl Sagan says in his book *Pale Blue Dot:* "Some planetary civilizations see their way through, place limits on what may and what must not be done, and safely pass through the time of perils. Others, not so lucky or so prudent, perish."

My own view is that we will successfully negotiate the hazards threatening our species. We will not kill ourselves off. We will not die off from disease. We will wax and wane in numbers as the long roll of time still facing this planet buffets our species with all manner of climate changes, asteroid impacts, runaway technology, and evil robots. We will persevere. But the animals and plants along for the ride on this planet that we have so cockily co-opted will not be so fortunate.

Perhaps this view that we are unkillable—at least as a species—is naïve. But even if we are to live as long as an *average* mammalian species—between 1 and 3 million years—we still have huge stretches of time left, for our species is barely a quarter of a million years old. And who says we are average? My bet is that we will stick around until the very end of planetary habitability for this already old Earth.

*Survivors at the twilight of the planet.*

TEN

# DEEP TIME, FAR FUTURE

> The Time Traveler (for so it will be convenient to speak of him) was expounding a recondite matter to us. His gray eyes shone and twinkled, and his usually pale face was flushed and animated. The fire burnt brightly, and the soft radiance of the incandescent lights in the lilies of silver caught the bubbles that flashed and passed in our glasses.
>
> —H. G. WELLS, *The Time Machine*

There is but one place that I know of where time is suspended.

High above the Coral Sea in a 747 bound for Australia, I sit with nose squashed against the frigid window, a bright moon and Southern Hemisphere stars starkly visible. All around me fellow passengers try to sleep, crammed into this silver cigar suspended above the Earth, chasing the night as we head west, the endless night. What time is it; what is time? A Quantas cabin attendant passes among the quiet throng, and unbidden, tells the passenger in front of me that there is no time up here, just distance. I shrug at this in affirmation; how can she be wrong, veteran of a thousand trans-Pacific voyages? It's 4 A.M. according to my watch, West Coast North American time, but 9 P.M. *here* according to my calculations, twelve hours into the flight. Three more hours to go to arrive in Australia, then three more hours on a cold airport bench, then two more in another plane to arrive in New Caledonia, my eventual destination. Is it past, present, or future here? Once again, H. G. Wells seems palpably present. On such a fight there seems no possible ending; all that came before is but memory, all to come is speculation. Reality is the cramped seat, the tiny window, the suspension above a dark Earth assumed to be below, and the baleful moon in its sprinkling of stars. I will never be closer to these stars, I muse. A book, quiet reflection, attempts at fitful sleep. The opposite of time.

And then, somehow, against all expectation, the flight *does* end, and time resumes, this voyage becomes memory, ending in a place where time, at least as counted by evolution, is suspended as well.

I had first come to New Caledonia in 1975, crossing the widest ocean for the first time, finally leaving the long den of school life. I had come to study an icon of evolution arrested, the chambered nautilus, the antithesis to this book—not the future of evolution, but an evolution ended. Then I was amphibious, for I had taken to the sea early in life. Donning scuba gear at 16, I became a salvage diver at 18 and an underwater instructor at 19. Being young (and thus immortal), I had no fear of the sea, for I felt more at home under its surface than I did living among the creatures of air. Thus, for three months in my twenty-fifth year of existence, I lived a life of the sea, of study, on an island that had once been part of Gondwanaland, splitting off during the Age of Dinosaurs to be carried by continental drift to its present tropical resting place, east of the Great Barrier Reef in Australia. New Caledonia became its own laboratory of evolution, eschewing mammals, instead evolving a unique fauna of birds and insects and flora derived from ancient Gondwanaland, a flora dating back to the time of the mammal-like reptiles of 250 million years ago. Cut off from the rest of the world, New Caledonia became a museum of the ancient, with fully half of its plants found nowhere else on Earth, and many of great antiquity. Evolution seems to have taken a long vacation here. In the first month of the new millennium, in my fiftieth year of existence, I returned to this ancient place, place of ancients.

Tropical, exuberant, gigantic as islands go, with high mountains and dark rock ripped from deep within the Earth's mantle and thrust upward in the cataclysmic breakup of Gondwanaland, New Caledonia looks like no other place on Earth. It is covered with forests of Norfolk pine and other relics of the Mesozoic Era, and only where humans have imported their plants and animals does it seem like the rest of the world. Coral reefs stretch out far from land, and one of the world's wonders, a Great Barrier Reef as impressive as that of nearby Australia, surrounds the three-hundred-mile-long island. It was outside of that reef that I had made the seminal discoveries that launched my scientific career. It was also in this same place that in my thirty-fifth year one of my closest friends would die in my arms following a dive together along this reef, our blood mingling as I vainly tried to breathe life back into his ruptured young body. Following that horrific extinction I had no desire to see this fatal shore again, but time eventually did its healing. It was to this same place that I finally returned, to the same stretch of reef where he had died, the blue sea finally purged of his red blood so long staining my memories of this place.

Twenty-five years seemed to have flashed by in an instant, and at the same time had crawled by so slowly. Old friends I had not seen for a decade and a half or more, friends who greeted me so warmly and with such emotion that I wondered who that man was that they had known, to be remembered so fondly, a man I no longer was? A huge block of time as measured by the life of a man, yet invisibly short as reckoned by the Earth's timekeeper, and by the clock of evolution. Yet as I walked the old beaches and places I had known so long ago, I found more changed on this island than not; I found that a quarter century of human development had radically transformed a place I remembered as still pristine. It was not just the new buildings, roads, factories, the vastly increased human population so evident everywhere, but the look of the place, the air less clear, the garbage now clogging beaches so pristine in my memory. I found myself yearning for the clear warm water of the reef, so far from shore, to see once again the ancient nautilus rising up from its deep daytime keep to stealthily prowl the shallows at night under the cover of darkness.

In this I was not disappointed.

Within a day of our arrival we were at sea, in a French oceanographic vessel chartered by the television crew that was behind the entire adventure. We anchored outside the barrier reef, 12 miles from shore, and started the chores necessary to find *Nautilus.*

The television company needed to get film of this ancient living fossil, and left nothing to chance. We set deepwater crab traps late one afternoon, a sure way to catch *Nautilus,* then spent the night waiting. At first light in the morning we winched the heavy traps up from their resting site 1,200 feet below. Seven nautilus and several temperamental deepwater eels and crabs were found in the traps, and we were ready to use these as stand-ins if, after all, we were unable to see wild animals in the night. With animals in the bank, there was nothing to do but wait for nightfall. Our boat was anchored right at the edge of the reef's drop-off, and I could see the change of color from light to dark blue as our boat drifted back and forth over the drop-off. I spent the day snorkeling in this ancient ecosystem—a habitat now so threatened in many parts of the globe.

The livid tropical sun ran its course, and the deep, rich colors of the reef created by the sun's equatorial light faded as the afternoon waned. Once again I felt the disorientation of a twilight running its course too quickly, the transition from day to night so rapidly accomplished in this region of the globe. In the deepening gloaming I donned old familiar gear, the museum pieces of equipment I had dived in so long ago in this place, well maintained and still as efficient as it had been a quarter century earlier. But it was not just my diving gear that seemed out of place. I was an anachronism among the younger divers with us on this expedition, young men and women sporting the colorful panache of the newest generation of diving equipment.

The warm, dark night of the tropics finally extinguished all light, and it was time to dive. We splashed in a bit after 8 P.M., immediately turning on bright underwater lights, and began searching for *Nautilus,* by now presumably newly ascended from its much deeper daytime haunts. I was diving with an old friend and fellow old man, Pierre Laboute, with whom I had seen my first wild nautilus almost exactly twenty-five years before. We brought with us two of the nautilus that we had caught earlier to be sure that the television company financing this elaborate and expensive expedition would get the footage they needed. But they proved unnecessary, and I surreptitiously released them back into their dark home, for near the end of the dive we saw not a trapped nautilus, but one that had come up to us from its retreat a thousand feet below to prowl these shuttered reef shallows in this dark tropical night. And so again two divers met in the dark, one a new species, the other one of the hoariest survivors of our planet's long history.

Once speared by our lights, the nautilus swam in long arcs, and we followed it for some time, two humans and a living fossil engaged in a slow-motion chase across the coral reef landscape in the dead of night. The white shell and magenta stripes of this nautilus seared brilliantly while transfixed by our powerful diving lights, and I have no doubt that the animal inside was terrified, if that word can be used for a creature with a very small brain. All afternoon the wind had been rising, and with it the sea, causing us to be buffeted even at the forty-foot depths we now swam through, but the nautilus simply motored through this swell, and we in our turn as well. Too quickly my air ran low, and my moments as a fish were over. When last seen, the nautilus was swimming in a seaward direction, back toward the security of the depths, and with this last vision Pierre Laboute and I headed back toward our boat in a sea now convulsing in the rising wind.

But my memory of that dive now turns not on this nautilus, but on time. Here we were encountering an animal not much different from the nautiloid cephalopods of 500 million years ago, a time when animals—any animals—were new things on this planet. In the nautilus it is not time, but evolution, that is suspended. And so, in that dark sea, I swam with joy at seeing this old friend, but with confusion at the unexpected strength of the feelings that this encounter was producing in me, and with the realization of how time had affected me as well: I could not dive as well, I was not as comfortable in the water, I had aged. There was a very good chance that a further twenty-five years would find me dead or, at seventy-five, certainly not diving the outer reaches of the New Caledonia barrier reef. Time, and time travelers.

I left the rich nighttime reef with its cargoes of animals and climbed up out of Mother Ocean, the clumsy tanks still slung on my back, into the face of a video camera, my face white and rubbery, nose running, a picture of a mechanically amphibious man caught in the web of time. I slung my mask and fins onto the pitching deck and smiled at my companions, sad for those who had to stay on the boat during our long dive, those that had missed a great privilege. They asked what we had seen, and I could only answer, wonders.

Coral reefs, with their diverse inhabitants, rich plankton, huge fish in great schools; indeed, these are wonders still here in this warm New Caledonian ocean. On the nearby land great flights of fruit bats swarm among the endemic tree species in the lush rainforests and dry forests, legacies of bygone geologic ages. Along the shores, spreading mangrove swamps guard a treasure trove of species.

Wonders, indeed.

Will those wonders continue?

All that has gone before in this book has given us a peek into the future, but that peek has been timid and so far limited to the near future as measured in thousands, or at most a paltry few millions, of years. But here, at the end, let us try a longer view. If the nautilus and its ilk can last 500 million years, persisting through trials of asteroid bombardment, tectonic cataclysm, rapid (and slow) climate change, reversal of the Earth's magnetic fields, nearby supernova explosions, gamma ray bursts, fluctuations in the intensity of the Earth's magnetic field, and surely much more still unknown to us, why not us? Why can't our species weather 500 million more years? Or a billion years, for that matter? Surely we can nudge aside the really large comets that head our way every million years or so.

To conclude this book, let us go forward in time until we reach that far-off land first seen by H. G. Wells. Let us go 500 million years into the future, the length of time that the nautilus has already existed, to a time closer to the end of the Earth than its beginning, and speculate about how the end of evolution—and of animal life on this planet—may come about.

By 500 million years from now the Earth, as a planet of life, will have aged considerably. Today, in this dawn of the Age of Humanity, we are already on a planet whose "habitability" has gone from middle to old age, a planet nearer the end of its life than the beginning. In those far future days, the engine of evolution will begin creating a rearguard action against the eventuality of our planet's death, a slow backing toward the final accounting that old age, even the Earth's, brings. By a billion years from now the Earth will no longer be habitable. Somewhere, then, between those two times will be a time when life on this Earth will have to adapt to ever-increasing heat and decreasing carbon dioxide. It is then, in that far future, that the types of animals and plants might finally prove to be exotic compared with our present-day biota.

The big problem, of course, will be the sun. Like all stars, it contains a finite amount of fuel, and as the tank empties, the temperature will increase. The amount of hydrogen being converted to helium will decrease, and heavier material will begin to accumulate. The sun will expand in size, and the Earth, the once equable Earth, will face the prospect of becoming the next Venus in our solar system: a desert without water, a place a searing heat, a burned cinder. That will be our fate. What will precede it?

Between 500 and 1,000 million years from now there will still be clinging survivors of the Cambrian Explosion of 500 million years ago, the last twigs of the once vigorous tree of life. Let us imagine a stroll along the seashore in such a world. The sun is gigantic, the heat searing. The equatorial regions are already too hot for all but microbial life, and it is only in the cooler polar latitudes that we can see the ends of animal life on Earth. Plant life is still present, but the amount of carbon dioxide in the atmosphere has shrunk to but a trace of its level during the first evolution of humanity. Only those plants evolved for life in this low-carbon-dioxide environment can be seen: low shrubbery with thick, waxy cuticles to withstand the searing heat and desiccation. There are no trees. Gone are the forests, grasslands, mangrove swamps, and meadows. The oceans are in the process of evaporating away, and huge salt flats now stretch for untold miles along their shores. There is no longer animal life in the sea, save for crustaceans adapted to the very high salt content. The fish are gone, as are most mollusks and other animals without efficient kidney systems, such as echinoderms, brachiopods, cnidarians, tunicates—all the groups that were never good at dealing with changes in the saltiness of the sea, or at moving into fresh water. There is still land life, for animals can be seen along the shores, but they are low, squat, heavily armored creatures, and their armor is not for protection from predation, but for protection from the ever-present heat, salt, and drying.

Inland from the sea there is a different vision. Lichen, a few squat low plants. Other desultory animals, some of them arthropods, a few of them vertebrates. All the rest of the world is a desert, a place of heat and dying.

The birds are gone. So too are the amphibians. Whole classes, even phyla, are now disappearing from the Earth like players from a stage when the play is ending.

There are still lizards, and snakes, and scorpions and cockroaches.

And humans.

All of humanity, or what is left of it, now lives underground in the cooler recesses of the Earth. It is as if at least part of H. G. Wells's vision has come true. In a sense, humans have become his Morlocks, a troglodyte species. There is too much radiation from the growing sun for humans to last long on the surface of the planet. Humanity, by necessity, has had to go underground, becoming the new ants of the planet. But physically, humans have not changed much. They know the end is near. There is no way off, no path to other, younger worlds. Space turned out to be too vast, the other planets in the solar system too inimical, the stars too far. Their Planet Earth is old and dying. They do not mourn the many animals the Earth once had. It is hard to remember things that happened 500 million years ago.

Once there was a future to evolution.

# BIBLIOGRAPHY

Adams, J., and F. Woodward. 1992. The past as key to the future: The use of paleoenvironmental understanding to predict the effects of man on the biosphere. *Advances in Ecological Research* 22: 257–309.

Akam, M., et al., eds. 1994. *The Evolution of Developmental Mechanisms.* Cambridge: Company of Biologists.

Alvarez, L., W. Alvarez, F. Asaro, and H. Michel. 1980. Extra-terrestrial cause for the Cretaceous-Tertiary extinction. *Science* 208: 1094–1108.

Benton, M. 1995. Diversification and extinction in the history of life. *Science* 268: 52–58.

Bourgeois, J. 1994. Tsunami deposits and the K/T boundary: A sedimentologist's perspective. Lunar Planetary Institute Cont. 825, p. 16.

Brown, L., C. Flavin, and H. French. 1999. *State of the World, 1999.* New York: Norton/Worldwatch Books.

Caldeira, K., and J. F. Kasting. 1992. The life span of the biosphere revisited. *Nature* 360: 721–723.

Caldeira, K., and J. F. Kasting. 1992. Susceptibility of the early Earth to irreversible glaciation caused by carbon ice clouds. *Nature* 359: 226–228.

Calvin, W. 1998. The great climate flip-flop. *Atlantic Monthly,* January, 47–54.

Carroll, S. B. 1995. Homeotic genes and the evolution of arthropods and chordates. *Nature* 376: 479–485.

Ceballos, G., and J. Brown. 1995. Global patterns of mammalian diversity, endemism and endangerment. *Conservation Biology* 9: 559–568.

Cohen, J. 1995. *How Many People Can the Earth Support?* New York: W. W. Norton.

Conway Morris, S. 1990. Late Precambrian and Cambrian soft-bodied faunas. *Annual Review of Earth and Planetary Sciences* 18: 101–122.

Conway Morris, S. 1992. Burgess Shale-type faunas in the context of the "Cambrian explosion": A review. *Journal of the Geological Society* (London) 149: 631–636.

Conway Morris, S. 1993. Ediacaran-like fossils in Cambrian Burgess Shale-type faunas of North America. *Palaeontology* 36: 593–635.

Conway Morris, S. 1993. The fossil record and the early evolution of the Metazoa. *Nature* 361: 219–225.

Covey, C., S. Thompson, P. Weissman, and M. Maccracken. 1994. Global climatic effects of atmospheric dust from an asteroid or comet impact on Earth. *Global and Planetary Change* 9: 263–273.

Davis, S., et al. 1986. *Plants in Danger: What Do We Know?* Gland, Switzerland: International Union for Conservation of Nature and Natural Resources.

DiSilvestro, R. 1989. *The Endangered Kingdom: The Struggle to Save America's Wildlife.* New York: John Wiley.

Dole, S. H. 1964. *Habitable Planets for Man.* New York: Blaisdell.

Dyson, G. B. 1997. *Darwin among the Machines: The Evolution of Global Intelligence.* Reading, MA: Addison-Wesley.

Ehrlich, P. 1987. Population biology, conservation biology, and the future of humanity. *Bioscience* 37: 757–763.

Eldredge, N. 1991. *The Miner's Canary: Unraveling the Mysteries of Extinction.* New York: Prentice Hall.

Ellis, J., and D. Schramm. 1995. Could a supernova explosion have caused a mass extinction? *Proceedings of the National Academy of Sciences U.S.A.* 92: 235–238.

Erwin. D. 1993. *The Great Paleozoic Crisis: Life and Death in the Permian.* New York: Columbia University Press.

Erwin, D. 1994. The Permo-Triassic extinction. *Nature* 367: 231–236.

Erwin, D. H. 1993. The origin of metazoan development. *Biological Journal of the Linnean Society* 50: 255–274.

Erwin, T. 1991. An evolutionary basis for conservation strategies. *Science* 253: 750–752.

Fuller, E. 1987. *Extinct Birds.* New York: Facts on File Publications.

Gagne, W. 1988. Conservation priorities in Hawaiian natural systems. *BioScience* 38: 264–271.

Garrett, L. 1994. *The Coming Plague: Newly Emerging Diseases in a World out of Balance.* New York: Penguin.

Gehrels, T., ed. 1994. *Hazards due to Comets and Asteroids.* Tucson: University of Arizona Press.

Gleiser, M. 1997. *The Dancing Universe: From Creation Myths to the Big Bang.* New York: Dutton.

Gott, J. 1993. Implications of the Copernican Principle for our future prospects. *Nature* 363: 315–319.

Goudie, A., and H. Viles. 1997. *The Earth Transformed.* Oxford: Blackwell.

Gould, F. 1991. The evolutionary potential of crop pests. *American Scientist* 79: 496–507.

Gould, S. 1994. The evolution of life on Earth. *Scientific American* 271: 85–91.

Gould, S. J. 1986. *Wonderful Life: The Burgess Shale and the Nature of History.* New York: W. W. Norton.

Gould, S. J. 1991. The disparity of the Burgess Shale arthropod fauna and the limits of cladistic analysis: Why we must strive to quantify morphospace. *Paleobiology* 17: 411–423.

Grayson, D. 1991. Late Pleistocene mammalian extinctions in North America: Taxonomy, chronology and explanations. *Journal of World Prehistory* 5: 19.

Gribbin, J. R. 1990. *Hothouse Earth: The Greenhouse Effect and Gaia.* New York: Grove Weidenfeld.

Groombridge, B., ed. 1992. *Global Biodiversity: Status of the Earth's Living Resources.* London: Chapman and Hall.

Grotzinger, J. P., S. A. Bowring, B. Saylor, and A. J. Kauffman. 1995. New biostratigraphic and geochronological constraints on early animal evolution. *Science* 270: 598–604.

Hallam, A. 1994. The earliest Triassic as an anoxic event, and its relationship to the End-Paleozoic mass extinction. In *Canadian Society of Petroleum Geologists,* Memoir 17, 797–804.

Hallam, A., and P. Wignall. 1997. *Mass Extinctions and Their Aftermath.* Oxford: Oxford University Press.

Hofman, P., A. Kaurfman, G. Halverson, and D. Schrag. 1998. A Neoproterozoic Snowball Earth. *Science* 281: 1342–1346.

Hsu, K., and J. Mckenzie. 1990. Carbon isotope anomalies at era boundaries: Global catastrophes and their ultimate cause. In *Global Catastrophes in Earth History,* edited by V. Sharpton and P. Ward, 61–70. Special Paper 247. Boulder, CO: Geological Society of America.

Isozaki, Y. 1994. Superanoxia across the Permo-Triassic boundary: Record in accreted deep-sea pelagic chert in Japan. In *Global Environments and Resources.* Canadian Society of Petroleum Geologists, Mem. 17, p. 805–812.

Jablonski, D. 1991. Extinctions: A paleontological perspective. *Science* 253: 754–757.

Jablonski, D. 1993. The tropics as a source of evolutionary novelty. *Nature* 364: 142–144.

Joy, W. 2000. Why the future does not need us. *Wired,* April, 238–262.

Kappen, C., and F. H. Ruddle. 1993. Evolution of a regulatory gene family: *HOM/Hox* genes. *Current Opinion in Genetics and Development* 3: 931–938.

Kathen, A. de. 1996. The impact of transgenic crop releases on biodiversity in developing countries. *Biotech. and Development Monitor,* no. 28: 10–14.

Kirchner, J. W., and A. Weil. 2000. Delayed biological recovery from extinctions throughout the fossil record. *Nature* 404: 177–180.

Kirschvink, J. A. 1992. Paleogeographic model for Vendian and Cambrian time. In *The Proterozoic Biosphere: A Multidisciplinary Study,* edited by J. W. Schopf, C. Klein, and D. Des Maris, 567–581. Cambridge: Cambridge University Press.

Knoll, A., R. Bambach, D. Canfield, and J. Grotzinger. 1996. Comparative Earth history and Late Permian mass extinction. *Science* 273: 452–457.

Kruess, A., and T. Tscharntke. 1994. Habitat fragmentation, species loss and biological control. *Science* 264: 1581–1584.

Kurzweil, R. 1999. *The Age of Spiritual Machines: When Computers Exceed Human Intelligence.* New York: Viking.

Lewis, J. S. 1999. *Comet and Asteroid Impact Hazards on a Populated Earth.* San Diego: Academic Press.

Lovelock, J. 1979. *Gaia: A New Look at Life on Earth.* Oxford: Oxford University Press.

Maher, K. A., and J. D. Stevenson. 1988. Impact frustration of the origin of life. *Nature* 331: 612–614.

Marshall, C., and P. Ward. 1996. Sudden and gradual molluscan extinctions in the latest Cretaceous of Western European Tethys. *Science* 274: 1360–1363.

May, R. 1988. How many species are there on Earth? *Science* 241: 1441–1449.

McKinney, M., ed. 1998. *Diversity Dynamics.* New York: Columbia University Press.

Morante, R. 1996. Permian and early Triassic isotopic records of carbon and strontium events in Australia and a scenario of events about the Permian-Triassic boundary. *Historical Geology* 11: 289–310.

Myers, N. 1985. The end of the lines. *Natural History* 94: 2–12.

Myers, N. 1993. Questions of mass extinction. *Biodiversity and Conservation* 2: 2–17.

Myers, N. 1996. The biodiversity crisis and the future of evolution. *The Environmentalist* 16: 124–136.

Pace, N. R. 1991. Origin of life: Facing up to the physical setting. *Cell* 65: 531–533.

Paine, R., M. Tegner, and E. Johnson. 1998. Compounded perturbations yield ecological surprises. *Ecosystem* 1: 535–545.

Pimm, S. 1991. *The Balance of Nature: Ecological Issues in the Conservation of Species and Communities.* Chicago: University of Chicago Press.

Pimm, S., G. Russell, J. Gittleman, and T. Brooks. 1995. The future of biodiversity. *Science* 269: 347–354.

Pope, K., A. Baines, A. Ocampo, and B. Ivanov. 1994. Impact winter and the Cretaceous Tertiary extinctions: Results of a Chicxulub asteroid impact model. *Earth and Planetary Science Express* 128: 719–725.

Raff, R. A. 1996. *The Shape of Life: Genes, Development, and the Evolution of Animal Form.* Chicago: University of Chicago Press.

Rampino, M., and K. Caldeira. 1993. Major episodes of geologic change: Correlation, time structure and possible causes. *Earth and Planetary Science Letters* 114: 215–227.

Raup, D. 1979. Size of the Permo-Triassic bottleneck and its evolutionary implications. *Science* 206: 217–218.

Raup, D. 1990. Impact as a general cause of extinction: A feasibility test. In *Global Catastrophes in Earth History,* edited by V. Sharpton and P. Ward, 27–32. Special Paper 247. Boulder, CO: Geological Society of America.

Raup, D. 1991. A kill curve for Phanerozoic marine species. *Paleobiology* 17: 37–48.

Raup, D., and J. Sepkoski. 1984. Periodicity of extinction in the geologic past *Proceedings of the National Academy of Sciences U.S.A.* 81: 801–805.

Raven, P. 1990. The politics of preserving biodiversity. *BioScience* 40: 769–774.

Retallack, G. 1995. Permian-Triassic crisis on land. *Science* 267: 77–80.

Sagan, C., and C. Chyba. 1997. The early faint sun paradox: Organic shielding of ultraviolet-labile greenhouse gases. *Science* 276: 1217–1221.

Salvadori, F. 1990. *Rare Animals of the World.* New York: Mallard Press.

Sepkoski, J. 1984. A model of Phanerozoic taxonomic diversity. *Paleobiology* 10: 246–267.

Schindewolf, O. 1963. Neokatastrophismus? *Zeitschrift der Deutschen Geologischen Gesellschaft* 114: 430–445.

Schwartzman, D., M. McMenamin, and T. Volk. 1993. Did surface temperatures constrain microbial evolution? *BioScience* 43: 390–393.

Sheehan, P., D. Fastovsky, G. Hoffman, C. Berghaus, and D. Gabriel. 1991. Sudden extinction of the dinosaurs: Latest Cretaceous, Upper Great Plains, U.S.A. *Science* 254: 835–839.

Sigurdsson, H., S. D'hondt, and S. Carey. 1992. The impact of the Cretaceous-Tertiary bolide on evaporite terrain and generation of major sulfuric acid aerosol. *Earth and Planetary Science Letters* 109: 543–559.

Sleep, N. H., K. J. Zahnle, J. F. Kasting, and H. J. Morowitz. 1989. Annihilation of ecosystems by large asteroid impacts on the Earth. *Nature* 342: 139–142.

Stanley, S. 1987. *Extinctions.* New York: W. H. Freeman and Company.

Stanley, S., and X. Yang. 1994. A double mass extinction at the end of the Paleozoic era. *Science* 266: 1340–1344.

Stenseth, N. 1984. The tropics: Cradle or museum? *Oikos* 43: 417–420.

Stone, C., and D. Stone, eds. 1989. *Conservation Biology in Hawaii.* Honolulu: University of Hawaii Press.

Stuart, C., and T. Stuart. A field guide to the larger animals of Africa. Cape Town: Struik Publishers.

Teichert, C. 1990. The end-Permian extinction. In *Extinction Events in Earth History,* edited by E. Kauffman and O. Walliser, 161–190. New York: Springer-Verlag.

Towe, K. M. 1994. Earth's early atmosphere: Constraints and opportunities for early evolution. In *Early Life on Earth,* edited by S. Bengston, 36–47. Nobel Symposium no. 84. New York: Columbia University Press.

Valentine, J. W. 1994. Late Precambrian bilaterians: Grades and clades. *Proceedings of the National Academy of Sciences U.S.A.* 91: 6751–6757.

Valentine, J. W., D. H. Erwin, and D. Jablonski. 1996. Developmental evolution of metazoan body plans: The fossil evidence. *Developmental Biology* 173: 373–381.

van Andel, T. H. 1985. *New Views on an Old Planet.* Cambridge: Cambridge University Press.

Vermeij, G. 1991. When biotas meet: Understanding biotic interchange. *Science* 253: 1099–1103.

Walker, J. C. G. 1977 *Evolution of the Atmosphere.* London: Macmillan.

Ward, P. 1990. The Cretaceous/Tertiary extinctions in the marine realm: A 1990 perspective: In *Global Catastrophes in Earth History,* edited by V. Sharpton and P. Ward, 425–432. Special Paper 247. Boulder, CO: Geological Society of America.

Ward, P. 1994. *The End of Evolution.* New York: Bantam Doubleday Dell.

Ward, P., and D. Brownlee. 2000. *Rare Earth: Why Complex Life Is Uncommon in the Universe.* New York: Copernicus (Springer Verlag).

Ward, P., and W. Kennedy. 1993. Maastrichtian ammonites from the Biscay region (France and Spain). *Journal of Paleontology,* Memoir 34 67.

Ward, P., W. J. Kennedy, K. MacLeod, and J. Mount. 1991. Ammonite and inoceramid bivalve extinction patterns in Cretaceous-Tertiary boundary sections of the Biscay Region (southwest France, northern Spain). *Geology* 19: 1181.

Ward, P. D. 1990. A review of Maastrichtian ammonite ranges. In *Global Catastrophes in Earth History,* edited by V. Sharpton and P. Ward, 519–530. Special Paper 247. Boulder, CO: Geological Society of America.

Whitmore, T. 1990. *An Introduction to Tropical Rain Forests.* Oxford: Oxford University Press.

Wilmer, P. 1990. *Invertebrate Relationships: Patterns in Animal Evolution.* Cambridge: Cambridge University Press.

Wilson, E. 1992. *The Diversity of Life.* Cambridge, MA: Harvard University Press.

Woese, C. R. 1987. Bacterial evolution. *Microbiological Reviews* 51: 221–271.

Woese, C. R., O. Kandler, and M. L. Wheelis. 1990. Towards a natural system of organisms: Proposal for the domains Archaea, Bacteria, and Eucarya. *Proceedings of the National Academy of Sciences U.S.A.* 87: 4576–4579.

# INDEX